The Foundations of Vacuum Coating Technology

Donald M. Mattox

Management Plus, Inc.
Albuquerque, New Mexico

NOYES PUBLICATIONS

WILLIAM ANDREW PUBLISHING
Norwich, New York, U.S.A.

Cover Art © 2003 by Brent Beckley / William Andrew Publishing

Library of Congress Catalog Card Number: 2003004260
ISBN: 0-8155-1495-6

Published in the United States of America by
Noyes / William Andrew Publishing
13 Eaton Avenue
Norwich, NY 13815
1-800-932-7045
www.williamandrew.com
www.knovel.com

This book may be purchased in quantity at discounts for education, business, or sales promotional use by contacting the Publisher.

Transferred to Digital Printing, 2010

Printed and bound in the United Kingdom

    Library of Congress Cataloging-in-Publication Data

Mattox, D. M.
    The foundations of vacuum coating technology / Donald M. Mattox.
        p. cm.
    ISBN 0-8155-1495-6 (alk. paper)
    1. Vacuum metallurgy--Congresses. 2. Metal coating--Congresses. I. Title.

    TN686.5.V3M38   2003
    669'.028'4--dc21
                        2003004260

William Andrew Publishing, 13 Eaton Avenue, Norwich, NY 13815 Tel: 607/337/5000 Fax: 607/337/5090

Dedication

The author would like to dedicate this book to his wife, Vivienne,
without whose support and encouragement this book would not have been written.
He would also like to dedicate the book to his granddaughter, Lindsey,
with the hope that when she grows up she will study the history
of those subjects that are of interest to her.

Acknowledgments

The author assumes responsibility for all errors of commission and omission and would like to encourage comments, corrections, and additions to be sent to donmattox@mpinm.com. The author would like to acknowledge Collin Alexander for providing material from the "proceedings" of the 1943 U.S. Army-sponsored conference on "Application of Metallic Fluoride Reflection Reducing Films to Optical Elements"—the text of which may be found on the SVC Web Site at www.svc.org; Ric Shimshock for the reference to the early work on magnetron sputtering on webs; Bob Cormia for his comments on the early development of planar magnetrons; and Angus Macleod for input on the early work in optical coatings. Russ Hill made some important contributions, as did Bob Waits, who called my attention to Fred Wehner's discussion on the etymology of the term "sputtering." The author would like to express thanks to Phil Baumeister for calling attention to the very important RLG patent and thus to some interesting patent litigation.

The author would particularly like to thank Julie Filatoff for her expert editing and formatting of this book, Linnea Dueker for her excellent job of scanning the illustrations, and Julie Romig for her critical reading of the manuscript. The SVC Oral History Project, located on the SVC Web Site at www.svc.org/H/H_OralInterviews.html, has also contributed some interesting details. Don Mattox would like to encourage anyone with historical papers, patents, pictures, or other materials related to the history of vacuum coating to contact him at donmattox@mpinm.com.

Preface

Vacuum coating processes consist of a source of vapor (atomic or molecular) of the material to be deposited, transport of the vapor to the surface to be coated, and condensation on the surface to form a thin film or coating with the desired properties. In the case of reactive deposition, availability of the reactive gas at the surface of the growing film and "activation" of the reactive gas are important.

Generally all the necessary ingredients for different vacuum-coating processes were present long before the process was used in production. In some cases, such as Edison's "Gold Moulded" cylinder phonograph record production (1902–1912), a vacuum coating process ("electrical evaporation," "cathodic disintegration" or, as it was later called, "sputtering") was the "enabling technology" that allowed mass production of the records. In other cases the use of vacuum coating processes has supplanted the use of other coating processes, such as electroplating, with a more environmentally friendly process. In still other cases the existence of vacuum coating processes allowed new applications and markets to be developed. An example is the deposition of "decorative-functional" coatings that have both a pleasing appearance (gold, brass, or gray-black for example) and wear resistance.

As with many ideas, inventions, and inventors, it is often difficult to establish the originator of any particular idea or invention. The search is complicated by the information being in both the patent literature and the scientific/technical literature. Patents seldom cite relevant technical papers and technical papers seldom cite relevant patent literature.

The field of vacuum coating is continually evolving and new processes, materials, and applications provide a fertile field for the historian.

"Those that do not know the history of their field are doomed to repeat its mistakes or end up in patent litigation." *(G. Santayana, paraphrased)*

Table of Contents

Introduction

Vacuum coatings processes use a vacuum (sub-atmospheric pressure) environment and an atomic or molecular condensable vapor source to deposit thin films and coatings. The vacuum environment is used not only to reduce gas particle density but also to limit gaseous contamination, establish partial pressures of inert and reactive gases, and control gas flow. The vapor source may be from a solid or liquid surface (physical vapor deposition—PVD), or from a chemical vapor precursor (chemical vapor deposition—CVD). The terms "physical vapor deposition" and "chemical vapor deposition" seem to have originated with C.F. Powell, J.H. Oxley, and J.M. Blocher, Jr., in their 1966 book _Vapor Deposition_ to differentiate between the types of vapor sources [1]. The term "vacuum deposition" is often used instead of PVD, particularly in the older literature such as Leslie Holland's classic 1956 book _Vacuum Deposition of Thin Films_ [2]. Vacuum coatings can also be formed by deposition of molten particles in a vacuum [3], but that process will not be covered here.

In PVD processing the vaporization may be from thermal heating of a solid (sublime) or liquid (evaporate) surface or by the non-thermal (momentum transfer) process of sputtering (physical sputtering). When the depositing material reacts with the ambient gaseous environment or a co-deposited material to form a compound, it is called reactive deposition. If the vapor source is from a solid molecular species (such as silicon oxide), some of the more volatile constituents may be lost upon vaporization. If these lost species are replaced by a reactive species from the deposition environment, the process may be called quasi-reactive deposition. The particle density in the vacuum may be such as to sustain a plasma discharge that provides ions and electrons. This plasma "activates" reactive gases for reactive deposition processes and aids in the decomposition of chemical vapor precursors ("plasma deposition" or plasma-enhanced chemical vapor deposition—PECVD). In some cases PVD and CVD processes are combined in the same chamber at the same time to deposit the material in a "hybrid process." For example, the deposition of titanium carbonitride (TiCxNy or Ti(CN)) may be performed using a hybrid process where the titanium may come from sputtering titanium; the nitrogen is from a gas and the carbon is from acetylene vapor. Alloys, mixtures, compounds and composite materials can be deposited using a single source of the desired material or multiple sources of the constituents.

In many instances the term "thin film" is used when discussing PVD vacuum deposits. This is because most early applications did not rely on the mechanical properties of the deposited material; they relied on the optical and electrical properties of thin deposits. In many recent applications the vacuum-deposited materials have been used for mechanical, electrical, and tribological applications, and the deposit thickness has become greater. The term "thick films" is not appropriate because that term is used for paint-on, fired-on coatings from slurries [4]. The term "metallurgical coating" is sometimes used but would not seem to be applicable to coatings that are not metallic. So the term "vacuum coating" is used here to refer to both thin films and thick deposits deposited by PVD or CVD in a sub-atmospheric (vacuum) gaseous environment. As an arbitrary delineation, the term "thin film" will generally be used for deposits less than about 0.5 microns (5,000 Ångstroms or 500 nanometers) in thickness. These types of vacuum coating processes are not to be confused with the "vacuum coating" process in the paint industry where excess paint is removed from a moving surface by a "vacuum."

The history of vacuum coating processes is closely associated with the history and the development of vacuum technology, electricity, magnetism, gaseous chemistry, plasma technology, thermal evaporation, arcing, and sputtering. "Pioneering" work (E-1) in these areas has led to many advances in vacuum coating technology.

Early Vacuum Science and Technology

(General References [5-9])

In about 1640, Otto von Guericke made the first piston-type vacuum pump (which he called "air pumps") patterned after the water pumps that had been used for many years to remove water from mines [10]. Figure 1 shows a woodcut picture of a mine being pumped out using several stages of water pumps. In 1654 von Guericke performed the famous "Magdeberg hemispheres" demonstration that really introduced the vacuum pump to the scientific world. In 1643 Torricelli demonstrated the mercury barometer based on the water manometer experiments of Berti (~1640). In 1662 Boyle formulated Boyle's Law, and in 1801 Dalton expressed the Law of Partial Pressures. Piston-type vacuum pumps came into widespread use, but vacuum experiments had to be continually pumped because of the poor vacuum seals available at that time.

In the early to mid-1800s, heat-moldable, electrically insulating materials included gutta-percha, a material made from the sap of a tree native to Southeast Asia (F. Montgomery brought this to Europe, 1843); the molded plastic made from shellac and sawdust (and later coal dust) (H. Peck & C. Halvorson, Norway, 1850); and "India rubber," which was made from rubber tree sap, sulfur, and gum shellac (G. Goodyear, 1851). L. Baekeland invented "Bakelite," the first totally synthetic thermosetting plastic, in 1907. Glass forming (ancient) and porcelain fabrication (W. Bottger, Germany, 1710) were well understood by the mid-1800s.

In 1857 H. Geissler invented the platinum-to-glass seal that allowed "sealed-off" vacuum tubes to be produced. This was a major advance in vacuum technology. Neoprene ("artificial") rubber was invented by DuPont in 1933, and molded seals of this material began to be used by the vacuum community in the late 1930s and replaced wax sealing. As late as the 1960s, finding and fixing vacuum leaks was a major part of vacuum technology and "Glyptol" (a paint—GE), "black wax" (Apiezon—W) and "elephant s—" (a putty) (Apiezon—Q) were common fixtures in the vacuum facility. Many of the early experiments used hydrogen flushing to augment vacuum pumping to reduce oxygen in the chamber. During WWII the Germans used phosphorus pentoxide, a desiccant, in the vacuum chamber to remove water vapor when depositing optical coatings.

In 1855 H. Geissler invented the mercury piston pump, which was widely used but left mercury vapor contamination in the vacuum (mercury has a vapor pressure of 1.2 mTorr at room temperature). Töpler improved the Geissler pump, and in 1865 Sprengel invented the mercury siphon pump, which greatly reduced the labor and attention required for vacuum pumping. In 1874 McLeod invented the

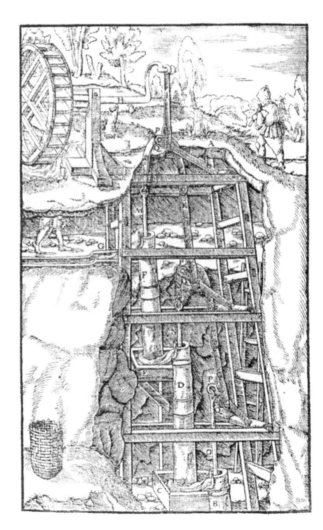

Figure 1. Multistage "syringe" water pumps removing water from a mine. From the book on mining, *De Re Metallica* (1556) [10].

"McLeod vacuum gauge," which was capable of measuring very low pressures. By using a combination of Geissler and Sprengel vacuum pumps, cold trapping and "flaming," good vacuums (about 10^{-6} atmospheres) could be achieved. Thomas Edison used this combination of techniques in the production of the first carbon-filament electric light bulbs in 1879. Figure 2 shows vacuum pumping, 1880s style. It was not until about 1895 that liquefied air was used to trap mercury vapor pumps. This was after T. Linde and J. Hampson had introduced "regenerative cooling" as a means of liquefying gases and J. Dewar invented a means for storing liquefied gases (Dewar flask). In 1901 H.C. Booth invented the first electrically-driven vacuum cleaner. The term "vacuum cleaner" was first used in Booth's advertisements.

In 1907 W. Gaede invented the oil-sealed rotary vane mechanical pump. By 1910, electric-motor-driven, oil-sealed, rotary vane pumps were in common use. In 1913 the mercury diffusion pump was invented by I. Langmuir and was improved by W. Gaede in 1922. In 1926 C.R. Burch replaced the mercury with low-vapor pressure oil, though well-trapped mercury diffusion pumps were considered the "cleanest" high-vacuum pumps for many years afterward. In 1937 L. Malter built the first metal oil diffusion pump; until that time fractionating and non-fractionating diffusion pumps were made of glass. The oil diffusion pump ("diff pump") has remained the principal high-vacuum pump used on many large vacuum coating systems. During WWII, the company Distillation Products, Inc. (later Consolidated Vacuum Corp.—CVC) and National Research Corporation (NRC—later a division of Varian Corp.) were the major suppliers of vacuum equipment to the vacuum coating industry.

Cryopumps and turbomolecular pumps have widespread use in the vacuum coating industry where small chambers are pumped or where oil contamination is a major concern. The cryopump developed from the use of cold traps in vacuum chambers and also cold shrouds in space simulators [11]. Liquid helium cooling developed after the invention of the "closed-loop" helium cryostat (Gifford-McMahon cycle) in 1960 but did not come into widespread use until 1980, when Varian Corp. began marketing cryopumps. W. Gaede invented a turbopump in 1912, but the modern vertical-axis, high-RPM turbopump

Figure 2. Vacuum pumping, 1880s style. This type of vacuum pumping system was used by T. Edison in the early manufacture of light bulbs.

was developed independently in 1958 by H.A. Steinhertz and by W. Becker. Pfeiffer Corp. began marketing turbopumps in 1958.

In the mid-1990s molecular drag stages were added to turbomolecular pumps. This technology allowed turbopumps to exhaust to higher pressures so diaphragm pumps could be used as backing pumps. This combination allowed very oil-free pumping systems ("dry pumps") to be developed for critical applications. The use of piston pumps also returned due to the need for clean, dry pumps. Ultra-clean vacuum components (and their packaging) and vacuum systems were developed for the semiconductor industry around the late 1960s [12].

In 1906 W. Voege invented the thermocouple vacuum gauge, and in 1909 M. von Pirani invented the Pirani gauge. That year the hot cathode ionization gauge was invented by Von Baeyer. The modern Bayard-Alpert hot cathode ionization gauge was invented by R.T. Bayard and D. Alpert in 1950. W. Sutherland (1897), I. Langmuir (1913), and J.W. Beams (1960) advanced the concept of various viscosity (molecular drag, spinning rotor) gauges. Beams' high-speed rotor work was the basis of the modern spinning rotor (molecular drag) gauge. The modern capacitance manometer gauge was invented by A.R. Olsen and L.L. Hirst in 1929 but did not become commercially available until 1961 when MKS Corporation began marketing it. F.M. Penning invented the magnetron cold cathode ionization gauge in 1937. In 1922 S. Dushman wrote the classic book *Production and Measurement of High Vacuum*. Helium leak detectors and many other advancements in vacuum technology were developed during WWII in support of the isotope separation project.

In the early years vacuum chambers were predominately single chambers, though Edison patented a dual-chamber vacuum system in the early 1900s to speed up production of light bulbs. With the advent of integrated circuit (IC) technology in the 1960s the need for high volume production of substrates having several layers deposited on them led to the development of several types of chambers [12a] including the open-ended [12b] and closed-ended [12c] "in-line" systems. Several systems were patented that had a central vacuum transfer chamber [12d]. A problem with these central-chamber designs was probably the complicated transfer mechanisms needed. In 1987 Applied Materials Inc. introduced the first commercial multiple-chamber "cluster tool" for semiconductor processing. This system was used for sequential deposition of doped and undoped PECVD films with an intermediate plasma-etching step [13]. These cluster tool systems later allowed very precise "pick-and-place" movement of wafers from a central cassette into processing chambers in a random sequence. Also in the 1960s coating on strip metal was being developed [13a]. This led to the fabrication of thin metal foils by depositing difficult to fabricate metals, such as beryllium, on a mandrel and then removing it in a continuous manner.

With the advent of reactive deposition and hybrid processing, the control of gas composition and mass flow has become an important aspect of vacuum engineering and technology. This includes partial pressure control and gas manifolding in processing chambers. Many vacuum measurement instruments cannot be used for measurement of total gas pressures in the presence of a plasma in the range of interest to many PVD processes (0.5 to 20 mTorr). Capacitance manometer gauges and spinning rotor gauges are commonly used in these applications. Differentially pumped mass spectrometers can be use to monitor and control partial pressures of gases. In 1984 a mass spectrometric feedback method of controlling the partial pressure of reactive gases in reactive sputter deposition was patented. Optical emission spectroscopy is also used to control the partial pressures of reactive gases in reactive sputter deposition. Optical emission has been used for many years to detect the "end-point" in plasma etching for semiconductor processing.

The use of vacuum equipment for deposition (and etching) processes also introduces problems associated with pumping and disposing of possibly toxic, flammable, and corrosive processing gases, as well as reactive gases used for *in situ* cleaning of the vacuum systems. Specialized vacuum equipment and *in situ* chamber plasma-etch-cleaning techniques have been developed to address these concerns.

An important aspect of vacuum coating processing is often how to create a vacuum rapidly and reduce the water vapor contamination quickly. In creating a "clean" vacuum the problem of water vapor moving from its point of origin to the vacuum pump by way of numerous "adsorption-desorption" events on surfaces is often the controlling factor in removing water vapor from the chamber [14].

The term "vacuum" should be used with caution because it means different things to different people. To some people the term vacuum means that the gas density is so low that the gaseous species do not affect the process or phenomena being studied. To others it means a sub-atmospheric gas density that should, or must, be controlled during the process or study in order to have reproducible results. For example, in Jim Lafferty's 1980 book *Vacuum Arcs* he makes a point of saying, "If there is an arc there is no vacuum and where there is a vacuum there is no arc" [15].

Early Electricity and Magnetism

(General References [16-18])

In 1600 W. Gilbert wrote the book _De Magnete (On The Magnet)_. This book is considered to be the first scientific publication. In 1672 Otto von Guericke (of vacuum pump fame) built the first electrostatic (frictional) electricity-generating machine, which used a rotating ball of sulfur. "Friction electricity" was used for entertainment in the early years (E-2). In 1732 Gray described the conduction of electricity, and about 1745 the air capacitor (Leyden jar) was invented by von Kleist, which allowed electricity to be stored. The invention of the Leyden jar is sometimes erroneously credited to Prof. van Musschenbroek at the University of Leyden in The Netherlands. In 1749 Benjamin Franklin introduced the concept of positive and negative electricity and the conservation of charge. Franklin also introduced the word "battery" for a bank of Leyden jars (E-3). In 1800 Alessandro Volta invented the electrolytic "voltaic pile" (later called a battery) based on the observations of "animal electricity" by Luigi Galvani (1791) and others (see Figure 3). The science of electrochemistry had its beginning at that time [19]— for example, the electrodeposition of copper (Cruikshank in 1800) and electrolysis, which allowed the separation of oxygen and hydrogen from water (an accidental discovery by Nicolson and Carlyle in 1800). Napoleon Bonaparte was immediately interested and supported the construction of very large arrays of batteries, as did others. For example, the Russians built an array of 4,200 Cu-Zn cells in 1803 at St. Petersburg's Medical and Surgical Academy. In 1810 Sir Humphry Davy produced a manmade arc between two electrodes. ("Arc" is from the word "arch," which is the shape of a long arc between two electrodes in air due to heating and convection.) Davy is generally credited with producing the first manmade arc, although a Russian, Vasilli V. Petrov, reported the same effect in 1803.

In 1820 H.C. Oersted detected the magnetic field around a current-carrying wire, and in 1821 Ampére invented the galvanometer. In 1831 Michael Faraday discovered electromagnetic induction. The first continuous generation of electricity, both AC and DC, using induction was developed by Hypolite Pixii in 1832. M. Nollet improved Pixii's design in 1849 or 1850. This led to the first commercial use of mechanically generated electricity. In 1836 N.J. Callan made the first induction coil to produce pulses of high voltages by periodically making and breaking a DC circuit. In 1851 H.D. Rühmkorff built a high-quality induction coil ("Rühmkorff coil") that allowed the generation of high voltages, and this device was widely used with gas-discharge tubes for many years.

In 1858 J. Plücker reported the bending of "cathode rays" by a magnetic field. In 1864 James Clerk Maxwell wrote the book _On a Dynamic Theory of Electromagnetic Fields_. In 1883 T. Edison placed a plate in one of his early lamp bulbs and noted that the current was higher when a filament, at a negative potential, was heated to incandescence ("Edison Effect") [20]. He was awarded a patent for the use of the changing current flow to "control machinery." In 1888 J. Elster and H. Geitel "rediscovered" the

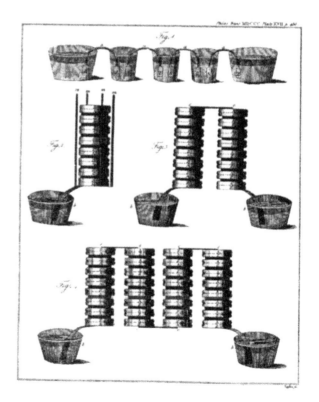

Figure 3. Four connections of "voltaic piles" (alternate silver and zinc discs separated by moist paper) as described in A. Volta's letter to Sir Joseph Banks of the Royal Society of London bearing the title, "On the Electricity excited by the mere Contact of conducting Substances of different kinds," dated March 20, 1800.

"Edison Effect," and they are often credited with discovering thermoelectron emission. In 1899 J.J. Thompson noted the same effect from a cathode of heated CaO and proposed "ionization by repeated contacts." In 1891 G.J. Stoney introduced the word "electron" for the charge passing through an electrolyte. In 1897 J.J. Thompson started his many studies on cathode rays that led to the identification of the electron (which he called a "corpuscle") in 1898. In 1895 Röntgen (or Rsntgen) discovered X-rays. In 1904 Fleming invented the "Fleming Valve," which was the forerunner of the triode vacuum tube.

Around the turn of the century there was an ongoing debate over which was better, direct current (DC) voltage advocated by Thomas Edison or alternating current (AC) voltage advocated by Nikola Tesla and George Westinghouse, who had purchased Tesla's patents (E-4, E-5). Of course AC won out partially because of its ability to step up or step down voltages using transformers. Until about the mid-1970s most high-voltage DC was produced using stepped-up voltages and electron tube rectifiers, such as the mercury vapor rectifiers. With the advent of solid-state electronics, high-voltage DC could be produced using solid-state rectifiers, but the electronics became more susceptible to arcing.

Radio frequency (RF) power is used to sputter dielectric materials. Until the mid-1970s crystal-controlled oscillators were used for frequency control (13.56 MHz); after that, solid-state oscillators were used. Around 1990, solid-state bipolar-pulsed power supplies became available with frequencies up to about 250 kHz (10 kHz to 250 kHz, "mid-frequency" AC) [21]. Pulsed power became an important option for sputtering and substrate biasing (E-6). High-power, mid-frequency AC power supplies could be produced at a much lower cost than comparable RF power supplies.

In 1928 Thompson discovered the diffraction of electrons as they passed though a thin film. In the late 1930s an "electron trap" was developed to confine electrons near a surface using a combination of electric and magnetic fields. This enhanced the plasma density near the surface and was called a "Penning Discharge." The Penning discharge was used to sputter from the inside of a cylinder (Penning 1936) and from a post (Penning and Mobius, 1940) This was a pioneering work in sputtering. The first device called a "magnetron" was invented by Albert W. Hull in about 1920. The cavity magnetron that led to the radar of WWII was invented by H.A. Boot and J.T. Randall in 1940 [22].

Early Plasma Physics and Chemistry

(General References [23-27])

In 1678 J. Picard noted a glow in the top of an agitated mercury barometer ("Picard's Glow"). Around 1720, F. Hawksbee used "frictional electricity" to generate a plasma in a vacuum that was intense enough "to read by." Scientists used "frictional electricity" to study the chemistry in electric sparks and plasmas before 1800. After the invention of the voltaic battery in 1800 by A. Volta, the study of the chemistry in electric arcs rapidly developed.

Since the mid-1800s there have been a number of studies of glow discharges and the spectral emission from the glows. The first glow (gas) discharge "vacuum tube" was made by M. Faraday in 1838 using brass electrodes and a vacuum of approximately 2 Torr. In 1857 Heinrich Geissler, who was the glassblower for Professor Julius Plücker, invented the platinum-to-glass seal that allowed sealed-off glow-discharge tubes (Geissler tubes) to be produced. In 1860 J.H. Hittorf, a pupil of Plücker's, noted that "cathode rays" (electrons from the cathode) projected "shadows" in a gas discharge tube (E-7). In 1885 Hittorf produced an externally excited plasma ("electrodeless ring discharge") by discharging a Leyden jar through a coil outside the glass chamber. W. Crookes made a number of studies using gas discharge tubes ("Crookes' tubes") (E-8). Figure 4 shows a modern rendition of the Crookes' tube [27]. J.J. Thompson made a number of studies which indicated that the cathode ray was composed of negative particles that were the same as Stoney's "electrons." In 1886 E. Goldstein, using a perforated cathode, identified the positively charged "proton" that was about 2,000 times heavier than the electron. In 1894 W. Ramsay discovered the inert gas argon.

In the mid-1920s I. Langmuir developed the small-area plasma probes that allowed characterization of plasmas (charge densities and particle temperatures). In 1926 Arthur R. von Hippel observed the optical emission spectra of sputtered atoms in a plasma. Although Hittorf had produced externally excited plasmas in 1885, the mechanism of this generation was in dispute until 1929 when McKinnon showed that at low gas densities the discharges were capacitively driven by coupling between the low- and high-voltage ends of the coil, while at high gas densities the plasma was inductively coupled between the turns of the coil. The inductively coupled plasma (ICP) was the basis of some early plasma sources. Reference [28] gives an example of a high-enthalpy ICP source. In the late 1930s an electron trap, which used a combination of electric and magnetic fields parallel to the surface called the "Penning effect," was used to enhance the plasmas near the surface in sputtering from cylindrical-hollow (inverted) magnetrons and cylindrical-post magnetrons. Work on these "surface magnetron" sputtering configurations was curtailed by WWII. The Penning effect was incorporated into a number of other applications such as vacuum gauges, sputter-ion pumps, and microwave tubes. Ions in the plasma are neutralized by contact with a surface. The time and rate of diffusion of ions to a surface determines the lifetime (decay time) of the plasma after the power to sustain the plasma has been cutoff. The lifetime of the plasma can be extended by using a magnetic field to confine the plasma away from surfaces.

The first "ion guns" were developed by NASA for space propulsion. Ion sources are of two types—an ion "gun," which has an extraction grid structure and produces an ion beam with a defined ion energy and low beam dispersion, and a "broad-beam" ion source, which produces an ion beam with a large dispersion and a spectrum of ion energies [29]. Research on power generation by nuclear fusion and studies of chemical synthesis in high-density plasmas accelerated the development of plasma confinement and the generation of high-density plasmas in the 1960s. These devices used electric and magnetic fields to confine and focus the plasma. Examples are the work of Gow and Ruby in 1959 [29a]. Also Filipova, Filipova, and Vingredov (1962) and Mather (1964) [30] made "pinched" (focused) plasmas. The

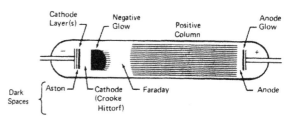

Figure 4. Glow discharge tube showing the various named regions of the discharge at low pressure [27].

general concept of containing electrons by the use of electric and magnetic fields, were used in the early 1970 to make "closed-loop" "surface magnetron" sputtering configurations.

The term "plasma," as we now use it, was proposed by I. Langmuir and L. Tonks of the General Electric Research Laboratories in 1928.

Some Scientific and Engineering Societies and Publications

The first scientific society in what became the USA was the American Philosophical Society, which was founded in 1743 under the instigation of Benjamin Franklin. The transactions of the American Philosophical Society began publication in 1771. Up to that time most American scientists published in European journals. In 1757 Benjamin Franklin was made a Member of the Royal Society (London) for his work in electricity [18, E-3]. In 1818 Benjamin Silliman, who has been called the "father of American scientific education," started publishing the American Journal of Science and Arts (E-9).

The American Vacuum Society (AVS) was formed in 1953 to provide a forum for those interested in the scientific and industrial development of vacuum equipment and processes. At first its primary interest was vacuum melting. It was not until about the Fifth Symposium (1957) of the AVS that papers on vacuum coating began to be presented. The AVS began publishing the *Journal of Vacuum Science and Technology* in 1964. In 1974 the AVS Conference on Structure/Property Relationships in Thick Films and Bulk Coatings was held in San Francisco. The meeting was sponsored by the Vacuum Metallurgy Division of the AVS, with Roitan (Ron) Bunshah as the Program Committee Chairman. This conference became an annual event called the International Conference on Metallurgical Coatings (ICMC), and later the ICMCTF when the AVS Thin Film Division became involved. The proceedings of these meetings are published in *Thin Solid Films* and *Surface and Coating Technology*. In 2001 the American Vacuum Society became AVS—The Science and Technology Society as its interests diverged from the use of the vacuum environment.

The Society of Vacuum Coaters (SVC) was formed in 1957 to provide a forum for developments in the industrial application of vacuum coatings. The papers that were presented at the first few conferences were primarily directed toward the decorative metallizing industry. The proceedings of these conferences are published by the SVC.

The Electrochemical Society was founded in 1902 and has published many articles related to vacuum coating technologies.

The International Union for Vacuum Science, Technique, and Application (IUVSTA), which is a society of organizations, was formed in 1962 upon the dissolution of the International Organization for Vacuum Science and Technology (IOVST), which was an organization that allowed individual and corporate members as well as organizations. In 1980 the IUVSTA organized around divisions, and the Thin Film Division was one of the first. The Thin Film Division developed from the International Thin Film Committee that was formed in 1968 by Klaus Behrndt [31].

Patents and the U.S. Patent Office

The existence of the U.S. Patent Office was specified by the Continental Congress in the U.S. Constitution. The first U.S. patent (#1) was issued in 1836. Until that time the patent process was very poorly organized. Up until 1880 the USPO asked for a model of the invention. Since 1880 the USPO can, but rarely does, ask for a model—an exception is a patent for a perpetual motion machine.

Patent litigation, with its high cost, has historically been used as a corporate "weapon" to spend (or threaten) people away from the use of patented work. Recently (Nov. 29, 2000) this may have changed due to a decision by the U.S. Court of Appeals in the case of *Festo Corp. v. Shoketsu Kinzoku Kogyo Kabushiki Ltd.* (234 F.3d 558, 56 USPQ2d 1865 [Fed. Cir. 2000]). This decision, which makes it harder to prove infringement on patented work, is on appeal to the U.S. Supreme Court. Basically this decision means, "Until the Supreme Court issues a decision, inventors will have to do their own extensive research before filing rather than relying on a patent examiner" ("Patent War Pending," *Amer. Bar Assoc. Journal,* Vol. 87(11), p. 28 [November 2001]).

One of the first major applications of this decision in the vacuum coating field was the dismissal of the Litton Industries patent infringement suit against Honeywell over the IBAD vacuum coating of ring laser gyro (RLG) mirrors. At one time a jury had awarded $1.2 billion to Litton and that award was under appeal when it was set aside completely by the same court that issued the Festo ruling [32].

Ɖeposition Ρrocesses

Sputter Ɖeposition

(General References [33, 34])

W.R. Grove was the first to study what came to be known as "sputtering" (and sputter deposition) in 1852 [35], although others had probably previously noted the effect while studying glow discharges. Grove sputtered from the tip of a wire held close to a highly polished silver surface (the type used for a daguerreotype) at a pressure of about 0.5 Torr, as shown in Figure 5. He noted a deposit on the silver surface when it was the anode of the circuit. The deposit had a ring structure. He made no studies on the properties of the deposited films since he was more interested in effects of voltage reversal on the discharge. In 1854 M. Faraday also reported film deposition by sputtering in a glow discharge tube. In 1858 Julius Plücker noted the formation of a platinum film inside of a discharge tube, creating a "beautiful metallic mirror" [36].

In 1877 Prof. A.W. Wright of Yale University published a paper in the *American Journal of Science and Arts* on the use of an "electrical deposition apparatus" to form mirrors and study their properties [37]. There is some confusion as to whether Wright was using sputtering or (gaseous) arcing [38], though it would seem that Wright was sputtering using an arrangement very similar to that of Grove (Figure 5) as shown in Figure 6 [39]. One major difference was that Wright used a swinging balance-pan fixture that allowed him to deposit ("paint") a

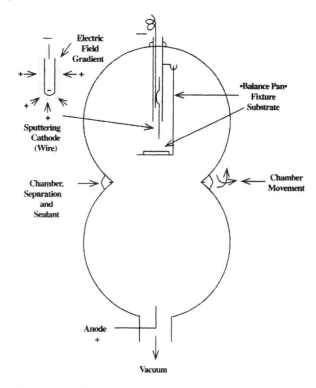

Figure 6. Wright's "electrical deposition apparatus" based on the description given in his paper (1877) [39].

film over a relatively large area rather than just at a point. Wright was not very specific in his description of the deposition process he used and the U.S. Patent Office used Wright's work as "prior art" when challenging part of T. Edison's 1884 patent application (granted 1894) on arc-based "vacuous deposition" [40]. Edison was granted his patent after maintaining that Wright's process used a "pulsed arc," whereas his was a "continuous arc." Edison also described Wright's work as a "laboratory curiosity." Professor Wright should be credited with being the first to characterize vacuum deposited films for their specific properties (color and reflectance).

In 1892 Edison used a "vacuous deposit" to "seed coat" his wax cylinder phonograph masters for subsequent electroplating [41]. In his 1902 patent on the subject [42] he indicated that the deposition process (arc deposition) described in his previous patent [40] wasn't suitable because of uniformity and heating problems (E-10), and in the figure in this patent (Figure 7) he showed a sputtering cathode (although he just called it an electrode) for depositing the metal. This process was used to make

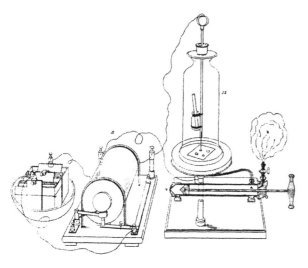

Figure 5. Grove's "sputtering" apparatus (1852) [35].

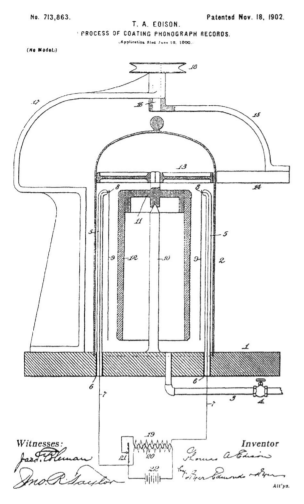

No. 713,863.   Patented Nov. 18, 1902.
                        T. A. EDISON.
            PROCESS OF COATING PHONOGRAPH RECORDS.
                    (Application filed June 10, 1900.)
(No Model.)

Witnesses: Inventor

Figure 7. Edison's arrangement for coating his wax phonograph masters using a sputtering cathode (9), a high-voltage induction coil, and an external magnetically driven (14) rotating fixture (13) (1902) [42]. Commercially this process was used to form Edison's "gold moulded" cylinder records (1902-1912).

the masters for Edison's "Gold Moulded Records" so-called because of the gold vapor given off by gold electrodes. These 2 minute records were introduced in January 1902. Edison should be credited with the first commercial use of sputter deposition. The "seed coat–electroplating" method is now used for depositing "seed-and-barrier" metallization on high-aspect-ratio surface features in semiconductor processing [42a]. In his 1894 patent Edison also referred to fabricating freestanding foils by stripping the deposit from the chamber walls. Edison also patented the use of the same sputter deposition equipment to form freestanding foils in 1915 [43]. He was interested in making filaments for his light bulbs using this

method of fabricating foils.

In 1891 W. Crookes published an article on sputtering and deposition, which he called "electrical evaporation," in the *Scientific American Supplement.* That article was probably the first "popular" publication on sputtering [44]. Crookes referred to Wright's work on producing mirrors. After the late 1800s, sputter deposition was used occasionally to make mirrors (other than silvered mirrors that were made by deposition of silver from a chemical solution) [45]. In 1912 Kohlschütter published the first comparison of the sputtering rates (sputtering yields) for various materials [46]. Later Almen and Bruce [47] made extensive studies on the sputtering yields of a number of materials. Collegon, Hicks, and Neokleous noted the variation of yield with initial target bombardment dose [48], a study that seems to rarely be noted in discussions on sputtering. In 1972 Kornelsen made several studies of gas incorporation into ion bombarded surfaces [49]. In 1963 van der Slice discussed the production of high-energy neutrals by charge-exchange processes in a glow discharge. This effect gives bombardment of the cathode, and nearby surfaces, by high-energy neutral species during sputtering [50].

The Bureau of Standards Circular #389 (1931) described the sputter deposition of mirror coatings [51]. The circular has some interesting information on the status of vacuum materials at that time; for example, "For joint grease a special solution of crude rubber and lard was made." In the early 1930s sputter deposition onto rolls of material (web coating or roll coating) was developed. The technique was first used to deposit silver on cloth in about 1931, in Leipzig, Germany. This was quickly followed by the deposition of gold and silver on glassine (waxed paper) by K. Kurz (Germany) and C. Whiley (England) for stamping foils [52]. By the late 1930s many applications of sputtering had been replaced by the thermal evaporation process that was developed in the early to mid-1930s. In 1933 the deposition of compounds by sputtering in a reactive gas ("reactive sputter deposition") was reported by Overbeck [53] for use as optical coating, though it was not widely used. The term "reactive sputtering" was introduced by Veszi in 1953 [54]. Reactive sputter deposition of tantalum nitride for thin film resistors was an early application. Tantalum nitride has about the same resistivity as tantalum and makes a more stable resistor material [55]. Previously sputter-deposited tantalum films had been electrolytically anodized to form a dielectric oxide, and with a sputter-deposited tantalum counter-electrode a thin film capacitor had been formed [56].

In the late 1930s the "crossed field" (electric and magnetic) electron trap was used to enhance plasmas in sputtering from cylindrical-hollow (inverted) magnetrons (Penning 1936) [57], and cylindrical-post magnetrons by Penning and Mobius in 1940 [58] in some pioneering work on sputtering. In the "Penning discharge" a combination of electric and magnetic fields is used to confine the plasma near the surface of the sputtering target. By having an appropriate magnetic-field configuration, the electrons are trapped near the surface, giving increased ionization and a short distance for the positive ions to be accelerated to the target. This allowed sputtering to be performed at lower pressures and lower voltages, and at higher rates than with non-magnetic DC sputtering. In the post- and hollow-cylinder sources, the magnetic field is parallel to the surface, and the electrons are reflected by "wings" or flanges on the ends. For example, a simple post becomes a "spool" shape. Figure 8 shows some of Penning's patent figures. Various forms of the Penning magnetrons have been developed since that time. Notable are the works of Penfold and Thornton on post cathode magnetron sputtering in the 1970s [59] and Mattox, Cuthrell, Peeples, and Dreike in the late 1980s [60]. Heisig, Goedicke, and Schiller [61] and Glocker [62] improved on the inverted magnetron configuration.

In 1962 Anderson, Mayer, and Wehner reported on the RF sputtering of a film that had been deposited on the inside of a glass window [63]. This cleaning technique was based on a suggestion by Wehner in 1955 [64] and the earlier observations of Robertson and Clapp on the sputtering of the inside of a glass tube when a high-frequency gas discharge was produced [65]. Davidse and Maissel pursued RF sputtering to produce films of dielectric material sputtered from a dielectric target in 1966 [66]. In 1968 Hohenstein used co-sputtering of a glass with RF, and metals (Al, Cu, Ni) with DC, to form cermet resistor films [67]. One aspect of the effectiveness of RF sputtering (>1 MHz) is due to the self-bias that is generated on the surface of the insulator [68]. RF sputter deposition did not have a ma-

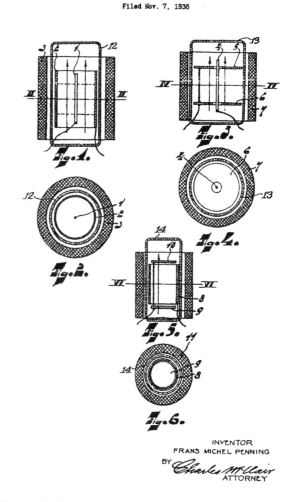

Figure 8. Penning magnetron configurations from his patent of 1939 [57].

jor impact on PVD processing for several reasons. These included the expense of large RF power supplies, and the problems associated with introducing high thermal inputs (i.e., high sputtering rates) into insulator materials without cracking them, since they are generally very brittle and poor thermal conductors. The one material that was widely RF sputtered was SiO_2, which has a low coefficient of thermal expansion (CTE). In 1967 Schick RF sputter-deposited a chromium coating on razor blades (Krona-Chrome™ razor) for corrosion protection.

In 1962 Wehner patented the process of deliberate concurrent bombardment "before and during" sputter deposition using a "bias sputter deposition" arrangement and mercury ions [69] to improve the epitaxial growth of silicon films on germanium sub-

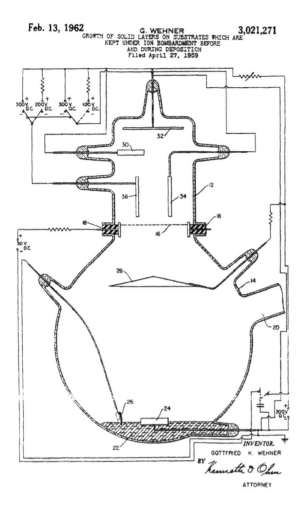

Figure 9. Wehner's Hg sputtering apparatus. In the figure, (36) is the sputtering cathode, (34) is the substrate, (22) is a mercury pool cathode, (16) is a graphite grid, (28) and (30) are anodes, and (32) is an "electron repeller" (1959) [69].

strates and to lower the "epitaxial temperature." Figure 9 shows the apparatus he used. Later this process became known as bias sputtering (bias sputter deposition) and is one form of "ion plating." Maissel and Schaible used bias sputter deposition to improve the purity of sputter-deposited chromium films [70]. Cooke, Covington, and Libsch used "getter" sputter deposition to improve the purity of sputter-deposited films [71]. In 1966 d'Heurle compared the use of bias sputter deposition, getter sputter deposition, and a combination of the two for preparing pure molybdenum films [72]. He obtained 30, 10, and 7 microhm-cm for the resistivities obtained with the respective techniques.

The triode sputtering configuration uses auxiliary plasma generated in front of the sputtering cath-ode by a thermoelectron emitting cathode and a magnetically confined plasma [73]. This arrangement was studied as a way to increase the plasma density and thus the sputtering flux that can be attained [74]. This sputtering technique lost its appeal with the development of magnetron sputtering.

In addition to the use of a deliberate substrate bias to accelerate charged particles to the substrate, high-energy particle bombardment can result from a "self-bias" on the substrate or from high-energy reflected neutrals resulting from ions that are de-ionized and reflected from the sputtering cathode [75]. This bombardment can cause "back sputtering" of some of the deposited material [76]. High-energy bombardment of the substrate can also be by negative ions (e.g., O⁻) accelerated away from the sputtering cathode [77]. Bombardment during deposition can also result in gas incorporation into the sputter-deposited material [78]. In non-biased, non-magnetron, planar diode sputter deposition, substrates in the line-of-sight of the cathode can be bombarded by high-energy electrons accelerated away from the cathode [79]. The substrates can be positioned in an "off-axis" geometry to avoid bombardment by positive ions, high-energy neutrals, or electrons [77].

In the early studies it was thought that the bombardment increased the purity of the sputter-deposited material and this was the reason that the deposited material had property values closer to those of the bulk (wrought) material [80]. It was not immediately recognized that the bombardment was causing densification of the deposited material; however, it was recognized that the bombardment affected crystallographic orientation [71, 81] and lowered the "epitaxial temperature" [69]. Bombardment was also shown to affect the film stress [82].

The effects of magnetic field on the trajectories of electrons had been realized even before Penning's work, and studies continued after Penning published his work on magnetrons [83]. The early Penning discharges used magnetic fields that were parallel to the sputtering target surface. Magnetron sources that use magnetic fields that emerge and reenter a surface ("magnetic tunnels") in a closed-loop pattern can be used to confine electrons near the surface is a closed pattern ("racetrack"). These confined electrons generate a high density plasma near the surface and were used in developing the "surface magnetron" sputtering configurations in the 1960s and early 1970s. These sources confine the electrons (and plasma) in a closed continuous "racetrack" on the target surface that does not cover the entire cathode surface. In 1962 Knauer patented an emerging and reentering closed-loop magnetic "tunnel" to trap electrons near the surface on both a post and a flat "washer-like" cathode electrode in a sputter-ion pump [84]. In 1963 Knauer and Stack described a closed magnetic tunnel on a post-cathode sputtering source for depositing films [85].

In 1968 Clarke developed a sputtering source using a magnetic tunnel on the inside of a cylindrical surface. This source became known as the "sputter gun" or "S-gun" [86] and is shown in Figure 10a. Mullaly at the Dow Chemical Co., Rocky Flats Plant (E-11), designed a magnetron source using a hemispherical target in 1969 [87]. Figure 10b shows the configuration of his magnetron sputtering source. Various magnetron configurations, including the planar magnetron, were patented by Corbani (patent filed July 1973, granted August 1975) [88]. Figure 10c shows figure 33 of Corbani's patent depicting a planar magnetron configuration. Chapin also developed a planar magnetron source (patent filed January 1974, granted August 1979) [89] and is credited with being the inventor of the planar magnetron sput-

tering source (E-12). Figure 10d shows the magnetron configuration depicted in Chapin's patent (F12).

Major advantages of these magnetron sputtering sources were that they could provide a long-lived, high-rate, large-area, low-temperature vaporization source that was capable of operating at a lower gas pressure and higher sputtering rates than non-magnetic sputtering sources. With their new performance characteristics, sputtering sources began to replace thermal evaporation in some applications and enabled new applications to develop. In 1975 sputter-deposited chromium was used on plastic auto grills (Chevrolet). In the late 1970s planar magnetron sputter deposition was applied to coating architectural glass. In 1977 Charoudi applied planar magnetrons to coating webs for window mirror applications under contract to the U.S. Department of Energy [90]. The first commercial wide-web sputter coating machine, made by Leybold, began operation at Southwall Corp. in 1980. In the mid-1970s reactively sputter-deposited hard coatings on tools began to be developed [91], and they became commercially available in the early 1980s.

One disadvantage of the early emerging/re-entry magnetic field magnetron sources was that the plasma was confined to a small volume near the surface of the sputtering target and thus was not avail-

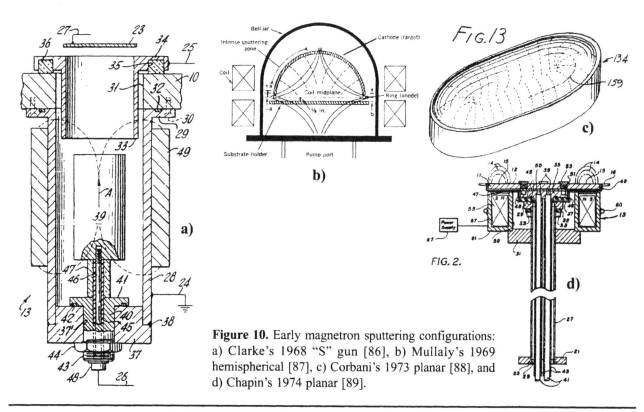

Figure 10. Early magnetron sputtering configurations: a) Clarke's 1968 "S" gun [86], b) Mullaly's 1969 hemispherical [87], c) Corbani's 1973 planar [88], and d) Chapin's 1974 planar [89].

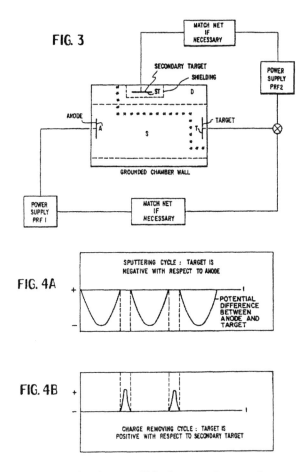

Figure 11. Circuit (a) and bipolar waveform on the target cathode (b) of Quazi's dual-cathode (one is "hidden") pulsed magnetron sputtering system (1986) [99].

able to provide "activation" of reactive gases near the substrate for reactive deposition processes nor ions for the bias sputtering process. This presented problems in the reactive deposition of compound films such as nitrides. However, this disadvantage could be overcome by the use of auxiliary plasma sources. These plasmas could be formed by providing RF as well as DC on the sputtering target or by having auxiliary plasma, often magnetically confined, near the substrate. The use of such auxiliary plasmas was cumbersome, and their use decreased with the advent of the concept of the deliberate "unbalanced" (UB) magnetron source by Windows and Savvides in 1986 [92]. The unbalanced magnetron allows some electrons to escape from the confining EXB field and create plasma in regions away from the target surface. If the escaping magnetic field is linked to other UB magnetron sources (N to S poles), the plasma-generation region can be significantly increased [93]. This technique is widely used today. In a 1972 patent

Davidse, Logan, and Maddocks used magnets above the substrate fixture to draw electrons to the substrate and create a "self-bias" on the substrate using a non-magnetron sputtering cathode [94].

The DC planar sputtering configuration became the most popular magnetron design. A disadvantage of the planar magnetron is that the stationary "race-track" erosion path gives non-uniform erosion over the target surface, low target-material utilization, and a non-uniform deposition pattern. In order to obtain uniform deposition on a substrate it is typically necessary to fixture and move the substrate in a specific manner or to have a moving magnetic field in the target. Another disadvantage is in reactive sputter deposition of highly insulating materials where the planar magnetron sources can be "poisoned" by the formation of a compound on the surface in areas outside the racetrack. This can cause surface charge buildup and arcing (flashover). To avoid this problem, RF can be superimposed on the DC target power, as was done by Vratny in 1967 on non-magnetron reactive sputter deposition [95], or "pulsed power" can be applied to the target. In 1977 Cormia patented the use of symmetrical or asymmetrical AC power on a single-cathode magnetron at a frequency of 400 Hz to 60 kHz ("mid-frequency") to create "pulsed power" on the cathode [96]. In this arrangement the target has a positive polarity during a portion of the wave cycle to allow electrons from the plasma to neutralize positive charge buildup on the poisoned surface.

Other target arrangements can be used to provide uniform erosion over the target surface. An example is to move the magnetic field behind the target. A variation on moving the magnetic field is to move the sputtering surface through the magnetic field by designing an elongated racetrack magnetic field formed by magnets inside a hollow cylinder that is rotated, which moves the surface through the field. This "rotatable cylindrical magnetron" was patented by McKelvey in 1982 [97]. The patent was assigned to Shatterproof Glass, but Airco bought the patent when Shatterproof Glass went bankrupt [98]. Major advantages to this design are the good utilization of the sputtering target material and the removal of most of the "poisoned" areas on the target as it rotates. This sputtering target design is in common use today (e.g., BOC's C-Mag™).

In 1987 Quazi patented sputtering from a single

magnetron sputtering target using discrete pulses of bipolar (alternately positive and negative) pulsed power that allowed the sputtering target to discharge any positive-charge buildup on the target (dielectric or "poisoned" metal surface) during the positive portion of the waveform [99]. Figure 11 shows the waveform and system configuration from his patent. This technique is similar to the "counterpulse" used to extinguish arcs in vacuum arc switch technology [100]. This concept was an improvement over the asymmetrical AC power used previously because the waveform could consist of square waves with fast rise and decay times and the pulse lengths could be varied independently. In 1993 Frach, Heisig, Gottfried, and Walde utilized the single-cathode bipolar pulsed power concept to reactively deposit dielectric films using a "hidden anode" to minimize the "disappearing anode" effect [101].

In 1988 Este and Westwood reported the mid-frequency dual-magnetron sputtering arrangement where the two targets were alternately cathodes and anodes [102]. This arrangement prevented the "disappearing anode" effect when reactively depositing highly insulating dielectric films and also prevented charge buildup and arcing (flashover) on the targets. Independently investigators at Kodak Research Laboratories were working on the same dual-cathode design and they used the technique to reactively sputter-deposit insulating films on web material in a roll coater in mid-1989 [103]. Scherer, Schmitt, Latz, and Schanz commercialized the dual planar magnetron configuration in 1992, as shown in Figure 12 [104], and it is in wide use today (e.g., Leybold Technologies' [later Applied Films'] TwinMag™).

In 2000 Glocker, Lindberg, and Woodard reported using dual inverted cylindrical magnetrons in a similar manner to dual-planar magnetrons [105]. In the early 1990s several companies began making high-power, mid-frequency, bipolar pulsed power supplies. Pulsed sputtering has been found to improve the properties of many films. This may be attributed to periodically higher plasma potentials caused by the pulsing, and thus higher sheath potentials; or it may be due to the high "spike" voltages during ignition, creating pulses of high-energy reflected neutrals that bombard the growing film.

There are many other variations on magnetron sputtering, including a target where the racetrack can either be a sputtering area or a steered arc track [106]. This system is used for the ArcBond™ process (Hauzer Techno Coatings) where the initial layer is deposited by arc vapor deposition, and the coating is built up by sputtering. This is done to increase adhesion and to minimize the number of molten particles ("macros") formed during arc vapor deposition that create defects in the coating. The magnetron plasma density has been enhanced by the injection of electrons from an auxiliary source such as a hollow cathode into the vicinity of the racetrack [106a].

In 1999 a magnetron system using one magnetron and two non-magnetron electrodes that were alternately at positive and negative potentials was introduced and called the "dual-anode magnetron," even though there was only one anode (and two cathodes) at a time present in the circuit [107]. The magnetron sputtering cathode is always negative with respect to the active anode and the polarity on the magnetron passes through zero twice on each volt-

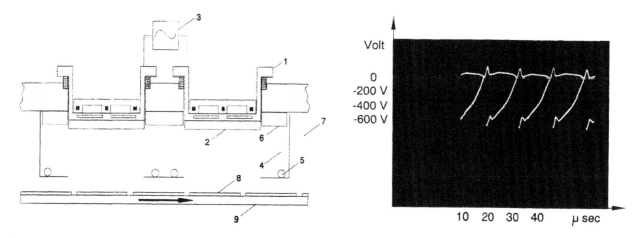

Figure 12. Scherer et al: a) dual planar cathode arrangement for coating flat glass; b) voltage waveform on a cathode (1992) [104].

age cycle, thus allowing charge buildup on the magnetron target to be neutralized. Anodes 1 and 2 are alternately positive (anodic) and negative with twice the negative voltage that appears on the magnetron. This allows the electrodes that act as anodes to be sputter cleaned on each cycle, thus avoiding the "disappearing anode" effect.

"Directed deposition" is confining the vapor flux to one axis by eliminating off-axis components of the flux. Directed deposition can be attained by collimation of the vaporized material. In sputtering this can be done using a "honeycomb" structure between the source and the substrate [108]. Minimum dispersion can also be attained by using a large separation between the source and substrate ("long-throw') and a low sputtering pressure [108a]. Another type of directed sputter deposition is the use of a gas flow to direct the sputtered material from the interior of a hollow cathode to the substrate [109]. This has been termed "gas flow sputtering" (GFS).

Generally very little sputtered material is ionized before it is deposited. Some techniques increase the ionization of the atoms by postvaporization ionization using plasmas and magnetic fields to form "film ions" that can be accelerated to the substrate surface when it has a negative bias (i-PVD) [110]. This encourages the depositing material to impinge on the substrate normal to the surface and increase the filling of high-aspect-ratio surface features. This is another type of "directed deposition" process.

Some work has been done using post-ionized sputtered species ("self-ions") to form the plasma used for sputtering ("self-sputtering"). Both a molten (thermal evaporating) cathode [111] and a solid magnetron target [112] with no gaseous species present have been used.

The technique of ion beam sputtering (and ion beam sputter deposition) was first used in the late 1960s [113]. Ion beam sources were developed from the ion propulsion engines developed by NASA in the 1950s [29]. There are two types of ion sources—an ion "gun" that has a defined ion energy and low beam dispersion, and a "broad-beam" that has a large dispersion and a spectrum of energies [114]. Two ion beams may be used—one to sputter the material and the other to bombard the depositing material. The use of ion beams to bombard the depositing material is called "ion beam assisted deposition" (IBAD). In this process, which will be discussed

under ion plating, the ion beam provides the bombardment of the depositing material from any vaporization source. The bombarding ions can be either an inert species or a reactive gas [115]. Ion beam sputter deposition is not widely used; however, ion beam "milling" (etching) to form surface features or ion beam "polishing" to smooth surfaces are used in some cases. One of the first significant commercial uses of ion beam reactive sputter deposition was the coating of ring laser gyro (RLG) mirrors [116]. The RLG has no moving parts and relies on extremely high-quality multilayer coated mirrors.

The scanning electron microscope (SEM) was first available in 1965 from Cambridge Scientific Instruments. The SEM provided an analytical tool that allowed studies on the growth morphology of sputter-deposited materials and the effects of concurrent bombardment on the growth [117]. In 1977 Thornton published a "structure-zone model," (SZM) patterned after the M-D diagram for evaporation-deposited coatings [118], which came to be known as the "Thornton Diagram" [119]. The Thornton Diagram illustrates the relationship between the deposit morphology, the deposition temperature, and the pressure in the sputtering chamber. Of course the sputtering pressure determines the flux and energy of the reflected high-energy neutrals from the sputtering cathode, so the diagram reflects the degree that the depositing material is bombarded by energetic particles during deposition. In 1984 Messier, Giri, and Roy further refined the structure zone model [120].

The columnar growth found in thick vacuum deposits is dependent upon the angle-of-incidence of the depositing flux, the surface roughness, the deposition temperature, and the amount of concurrent energetic particle bombardment [121]. By using a low angle-of-incidence and varying the angle-of-incidence (GLAD—glancing angle deposition), "sculpted films" can be grown [122].

Also in the 1970s there were a number of studies on the effect of processing variables, particularly concurrent bombardment, on the intrinsic stress of sputter-deposited films [123]. In 1988 periodic variation of pressure ("pressure cycling") was used by Mattox, Cuthrell, Peeples, and Dreike to control the total stress in thick sputter-deposited layers of molybdenum by depositing alternate layers having compressive and tensile stress. This was accomplished by periodically changing the gas pressure, and thus the flux and energy of bombardment by reflected high-energy neutrals [124].

In 1984 Sproul and Tomashek patented a mass spectrometric-feedback method of controlling the partial pressure of reactive gases in reactive sputter deposition [125]. Optical emission spectroscopy (OES) is also used to control the partial pressures of reactive gases in reactive sputter deposition [126]. OES developed from the "end-point" detection used in plasma etching [127].

With the advent of magnetron sputtering, controlled reactive sputter deposition, and the use of controlled concurrent ion bombardment as a process parameter, sputter deposition rapidly developed after the mid-1970s. Applications of sputter deposition have rapidly increased since that time. Today applications of sputter deposition range from depositing semiconductor device metallization, to coating architectural window glass for energy conservation, to coating tool bits, to coating plumbing fixtures for decorative purposes, to coating web material with transparent vapor barriers for packaging. In some cases sputter deposition has replaced electroplating or has displaced thermal evaporation PVD, but generally it has generated new applications and markets.

The term "sputtering" is probably an outgrowth of the controversy over whether this type of vaporization was thermal evaporation ("electrical evaporation") [44] or whether it was due to a non-thermal (momentum transfer) process [128]. Initially the English term "spluttering" (E-13) was used, which became sputtering several years later. To quote G.K. Wehner [129]:

> "Sometimes the question is raised as to who introduced the word 'sputtering.' A literature search shows that Sir J.J. Thompson [170] used the word 'spluttering' but that I. Langmuir and K.H. Kingdon eliminated the 'l' in their publications in the years 1920 to 1923. In 1923, the Research Staff of the General Electric Co., London [171] still used only the term 'cathode disintegration.' Kay [172] following a suggestion by Guentherschulze [173], tried without much success to introduce the term 'impact evaporation.'"

Unfortunately Wehner has the wrong date (1921) in his reference to Sir J.J. Thompson's book. The correct date for that publication is 1913. The manner in which Thompson used the term spluttering (e.g., "A well-known instance of this is the 'spluttering' of the cathode in a vacuum tube; —") [130] would seem to indicate that this was still not the first use of the term.

By 1930 the term "sputtering" was being used for the deposition process (e.g., "sputtered films") as well as the vaporization process. This at times leads to confusion. A more proper term for the deposition process would be "deposition by sputtering" or "sputter deposition."

Thermal Evaporation

(General References [2, 131-134])

Thermal evaporation is the vaporization of a material by heating to a temperature such that the vapor pressure becomes appreciable and atoms or molecules are lost from the surface in a vacuum. Vaporization may be from a liquid surface (i.e., above the melting point) or from a solid surface (i.e., sublimes). The author has arbitrarily defined 10^{-2} Torr as the equilibrium vapor pressure above which the free-surface vaporization in PVD-type vacuums is enough to allow vacuum deposition to occur at a reasonable rate [134]. If the material is solid at that temperature the material is said to sublime (e.g., Cr, Mg) and, if molten, is said to evaporate (e.g., Al, Pb, Sn, Mo, W). A few materials have a vapor pressure such that they can either be sublimed or evaporated at temperatures near their melting point (e.g., Ti). Some compound materials sublime and some evaporate.

Thermal evaporation studies in vacuum began in the late 1800s with the work of H. Hertz [135] and S. Stefan [136], who determined equilibrium vapor pressures—but they did not use the vapor to form films. Knudsen proposed "Knudsen's Cosine Law of Distribution" for vapor from a point source in 1909. In 1915 Knudsen refined the free-surface vaporization rate as a function of equilibrium vapor pressures and ambient pressure [137], and the resulting equation is known as the Hertz-Knudsen equation for free-surface vaporization. Honig summarized the equilibrium vapor pressure data in 1957 [138].

Thermal evaporation by "heating to incandescence" and film deposition was covered by Edison's 1894 patent (applied for in 1884) [40]. Edison did not mention evaporation from a molten material in his patent and many materials will not vaporize at an appreciable rate until they are at or above their melting temperature. Edison did not use the process in any application, presumably because radiant heating from the source was detrimental to the vacuum materials available at that time. In 1887 Nahrwold reported the formation of films of platinum by sublimation in vacuum [139], and he is sometimes credited with the first use of thermal vaporization to form films in a vacuum. In 1917 Stuhlmann reported depositing silver from an incandescent silver wire passed over a surface in vacuum to form a mirror [140]. In 1907 Soddy proposed the vaporization of calcium onto a surface as a method of reducing the residual pressure in a sealed tube ("gettering") [141]. This would have been the first "reactive deposition" process.

In 1912 von Pohl and Pringsheim reported forming films by evaporating materials in a vacuum from a magnesia crucible that was heated by a resistively heated foil surrounding the crucible, as shown in Figure 13 [142, E-14]. They are sometimes credited with the first deposition by thermal evaporation in vacuum. Langmuir studied the vaporization rate of materials in vacuum in 1913 [143] and reported forming films. In 1931 (possibly 1928, see E-15) Ritschl reported thermal evaporation of silver from

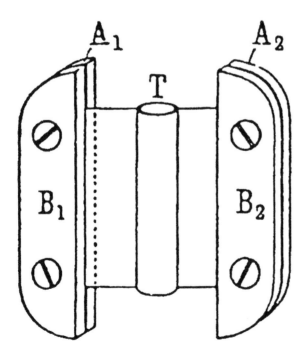

Figure 13. von Pohl and Pringsheim crucible evaporation source using a heated foil held in clamps (1912) [142, E-14].

a tungsten wire basket to form half-silvered mirrors [144]. Ritschl is often credited with being the first to use evaporation from a filament to form a film in vacuum. Cartwright and Strong reported evaporating metals from a tungsten wire basket in 1931 [145], but the technique was not successful for evaporating aluminum because molten aluminum wets and alloys with the tungsten wire, which causes it to "burn out" when there is a relatively large volume of molten aluminum. Aluminum was successfully evaporated by Strong in 1933 from heavy-gauge tungsten wire that was wetted by the molten aluminum [146]. John Strong began work on evaporated films in 1930 when he was trying to form reflective film onto alkali halide prisms for use in IR spectrometry at the University of Michigan.

In 1931 the Bureau of Standards in its circular on making mirrors [51] discussed the "common" technique of sputtering and "the deposition of metals by evaporation." In the description the Bureau stated, "This method of deposition ('thermal evaporation') has not been widely tested, and its possibilities are therefore little known, but it would seem to be especially valuable for small work where films of any readily volatile substance are required." Figure 14 (figure 4 in the circular) illustrates evaporation of material from an open wire coil in a glass chamber that has to be "cracked" in order to remove the deposited film. The chamber could be re-used by joining the sections with sealing wax. In the circular, W.W. Nicholas was named as the investigator at the Bureau of Standards and refers to other investigators (J. E. Henderson and A.H. Pfund) who were investigating thermal evaporation from filaments as a means of depositing films in vacuum. In that circular it was noted that J.E. Henderson was evaporating nickel that had been electroplated onto a tungsten wire to provide an "especially uniform deposit." Strong described the same technique in 1932 [147]. Edison, in his 1894 patent, discussed depositing (e.g., electroplating) a material on carbon, then heating the carbon to vaporize the material in vacuum [40].

J. Strong of Johns Hopkins University and later the California Institute of Technology (1947), where he worked on metallizing the 200" Palomar telescope, did a great deal of development using thermal evaporation of aluminum from multiple tungsten filaments for coating astronomical mirrors [148] (E-16, E-16a). Strong, with the help of designer

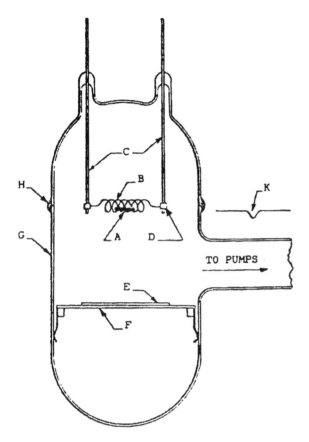

Figure 14. Glass vacuum chamber used for filament thermal evaporation. The chamber could be "cracked" (at H) and reused using sealing wax (1931) [51].

Bruce Rule, aluminum coated the 200" Palomar ("Hale") astronomical telescope mirror in 1947 (the telescope mirror was started before WWII) [149] using multiple (350) filaments and a 19 foot diameter vacuum chamber [150]. In 1937 D. Wright of GE began development of the sealed-beam headlight, which first appeared on autos in 1940 [151] (E-17).

In 1817 Fraunhoffer noted that optical lenses improved with age due to the formation of a surface film. Following this discovery many investigators artificially aged lenses to form antireflection (AR) coatings. For example, in 1904 H.D. Taylor patented (British) an acid treatment of a glass surface in order to lower the index of refraction and the reflectivity by producing a porous surface. In 1933 A.H. Pfund vacuum-deposited the first single-layer (AR) coating (ZnS) while reporting on making beam-splitters [152] and Bauer mentioned AR coatings in his work on the properties of alkali halides [153]. In 1935, based on Bauer's observation, A. Smakula of the Zeiss Company developed and patented AR coat-

ings on camera lenses [154]. The patent was immediately classified as a military secret and not revealed until 1940 (E-18). In 1936 Strong reported depositing AR coatings on glass [155]. In 1939 Cartwright and Turner deposited the first two-layer AR coatings [156]. Monarch Culter seems to be the first person to calculate the effect of multilayer coatings on optical properties. He did this as a Master's Degree thesis project [157]. This work preceded the work of Cartwright and Turner and, though unpublished, possibly inspired their efforts. One of the first major uses of coated lenses was on the projection lenses for the movie *Gone With the Wind*, which opened in December 1939 (S. Peterson of Bausch & Lomb Optical Co.) [158]. The AR coated lenses gained importance in WWII for their light-gathering ability in such instruments as rangefinders and the Norden Bombsight [159]. During WWII, baking of MgF_2 films to increase their durability was developed by D.A. Lyon of the U.S. Naval Gun Factory [160]. The baking step required that the lens makers coat the lens elements prior to assembly into compound lenses. In 1943 (Oct.) the U.S. Army (The Optical Instrument Committee, Frankfort Arsenal) sponsored a conference on "Application of Metallic Fluoride Reflection Reducing Films to Optical Elements." The conference had about 132 attendees. The proceedings of this conference (112 pages) is probably the first extensive publication on coating optical elements [161]. O.S. Heavens published his classic work, *Optical Properties of Thin Solid Films* (Butterworth Scientific Publications), in 1955.

The Germans deposited CaF_2 and MgF_2 AR coatings during WWII [162]. Plasma cleaning of glass surfaces is reported to have been used by Bauer at the Zeiss Company in 1934 [162]. The Schott Company (Germany) was also reported to have deposited three-layer AR coatings by flame-pyrolysis CVD during WWII [162].

Vacuum evaporation of metals (Cd and Zn) on paper web for paper-foil capacitors was begun in about 1935 by R. Bosch of the Bosch Company of Germany, who discovered that there was a "self-healing" effect when there was an arc between the low-melting-point thin film electrode materials [163]. By 1937 the Germans had demonstrated that the use of a "nucleating layer" increased the adhesion of zinc to a paper surface [164]. The effect of a nucleating layer on film formation had been noted by Langmuir as early as 1917 [165]. In 1958 the U.S. military formally approved the use of "vacuum cadmium plating" (VacCad) for application as corrosion protection on high-strength steel to avoid the hydrogen embrittlement associated with electroplated cadmium [166]. In recent years PVD processing has been used to replace electroplating in a number of applications to avoid the water pollution associated with electroplating.

In 1907 von Pirani patented the vacuum melting of refractory materials using focused cathode rays (electrons) from a glow discharge [167]. (W. Crookes noted that he could fuse the platinum anode in his gas discharge tubes in 1879.) In his patent von Pirani did not mention film formation, but he may have deposited films during the vacuum melting operation at low gas pressures. Figure 15 shows the figure from his patent that illustrates focused electron beam melting. In 1933 H.M. O'Brian and H.W.B. Skinner utilized accelerated electrons to heat a graphite crucible and evaporate materials [168]. Subsequently a number of non-focused ("work accelerated") electron bombardment evaporation source designs were developed. In 1951 Holland patented the use of accelerated electrons to melt and evaporate the tip of a wire ("pendant drop"), which involved no filament or crucible [169]. In 1949 Pierce described the "long-focus" electron beam gun for melting and evaporation in vacuum [170]. The long-focus gun suffers from shorting due to the deposition of evaporated material on the filament insulators that are in line-of-sight of the evaporating material. Deposition rates as high as 50 µm/s have been reported using e-beam evaporation [171].

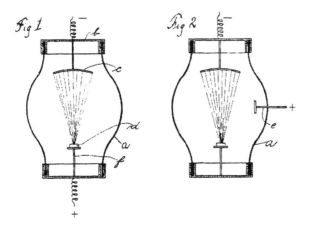

Figure 15. Vacuum melting using focused "cathode rays" (Pirani, 1907) [167].

To avoid exposure of the filament to the vapor flux, bent-beam electron evaporators were developed [172]. In 1968 Hanks filed a patent on a 270° bent-beam electron beam evaporation source [173] that has become the most widely used design. Rastering the electron beam allows the energy of the electron beam to be distributed over the surface. In 1978 H.R. Smith described a unique horizontally emitting electron beam (EB) vapor source [174]. The source used a rotating crucible to retain the molten material, and its function was to coat large vertical glass plates. As early as 1970 Kurz was using an electron-beam system to evaporate gold for web coating [52]. In electron beam evaporation a high negative "self-bias" can be generated on the surface of an insulating material or on an electrically isolated fixture. This bias can result in high-energy ion bombardment of the self-biased surface [175].

A number of thermoelectron-emitter e-beam source designs followed, including rod-fed sources and "multi-pocket" sources. The high voltage on the filament prevented the source from being used in a plasma where ions accelerated to the cathodic filament; this caused rapid sputter-erosion of the filament. In 1971 Chambers and Carmichael [176] avoided that problem by having the beam pass through a small hole in a thin sheet in a section of a plate that separated the deposition chamber from the chamber where the filament was located. This allowed a plasma to be formed in the deposition chamber while the filament chamber was kept under a good vacuum. The plasma in the deposition chamber allowed ion bombardment of the depositing film material as well as "activation" of reactive gas.

The use of a hollow cathode electron emitter for e-beam evaporation was reported by J.R. Morley and H. Smith in 1972 [177]. In 1951 Picard and Joy described the use of evaporation of materials from an RF-heated crucible [178]. In 1966 Ames, Kaplan, and Roland reported the development of an electrically conductive TiB_2/BN composite ceramic (Union Carbide Co., UCAR™) crucible material that was compatible with molten aluminum [179].

Directed deposition is confining the vapor flux to one axis by eliminating off-axis components of the flux. Directed deposition can be attained by collimation of the vaporized material. In evaporation this was done by Hibi (1952), who positioned a tube between the source and the substrate [180]. Collimation was also attained by H. Fuchs and H. Gleiter in their studies of the effects of atom velocity on film formation using a rotating, spiral-groove, velocity selector [181]. In 1983 Ney described a source that emitted a gold atom beam with a 2° divergence [182]. Recently "directed deposition" has been obtained using a flux of thermal evaporated material projected into a directed gas flow [183].

When thermally evaporating alloys, the material is vaporized with a composition in accordance to Raoult's Law (1887) [184]. This means that the deposited film will have a continuously varying composition unless very strict conditions are met as to the volume of the molten pool using a replenishing source [185]. One way of avoiding the problem is by "flash evaporation" of small volumes of material. In 1948 L. Harris and B.M. Siegel reported flash evaporation by dropping small amounts of material on a very hot surface so that all of the material was vaporized before the next material arrived on the hot surface [186]. In 1964 Smith and Hunt described a method for depositing continuous strips of alloy foils by evaporation [187]. Other free-standing thin film structures are also deposited, such as beryllium X-ray windows and nuclear targets [188].

Evaporation or sublimation of compounds can result in extensive molecular disassociation. Some compound materials can be vaporized without significant disassociation. These include many halides, sulfides, and selenides, as well as a few oxides (such as SiO). Many of these compound materials were used in early optical coating "stacks," and for many years thermal evaporation was almost the only PVD technique for depositing optical coatings. During sublimation of these materials, some of the material comes off as "clusters" of atoms (e.g., Se) or molecules (e.g., SiO). "Baffle" or "optically dense" sources were developed that required vaporization from several hot surfaces or deflection of the particles before the vapor could leave the source [189]. This generated a more uniform molecular vapor. Baffle sources can also be used to evaporate material in a downward direction. Sublimed SiO coatings were used on mirrors for abrasion resistance by Heraeus (Germany) before WWII [190]. Following the deposition they were heated in air to increase oxidation. In 1950 G. Hass evaporated lower-oxide materials in an oxygen atmosphere in order to increase the state of oxidation [191].

In 1952 Aüwarter patented the evaporation of metals in a reactive gas to form films of compound materials [192]. In 1960 Aüwarter proposed that evaporation of a material through a plasma containing a reactive species be used to form a film of compound material [193]. Many investigators studied these methods of depositing transparent optical coatings. In 1964 Cox, G. Hass, and Ramsey reported their use of "reactive evaporation" for coating surfaces on satellites [194]. In 1972 R.F. Bunshah introduced the term "activated reactive evaporation" (ARE) for evaporation into a reactive plasma to form a coating of a compound material [195].

"Gas evaporation" is a term used for evaporation of material in a gas pressure high enough to result in multi-body collisions and gas phase nucleation. This results in the formation of fine particles that are then deposited [196]. A.H. Pfund studied the optical properties of fine particles in 1933 [197]. It is interesting to note that during gas evaporation performed in a plasma, the particles become negatively charged and remain suspended in the plasma—since all the surfaces in contact with the plasma are negative with respect to the plasma. Electrically charged gas-phase-nucleated particles can be accelerated to high kinetic energies in an electric field. This is the basis for the "ionized cluster beam" (ICB deposition process introduced by Takagi et al in 1974 that used a thermal-evaporation, nozzle-expansion type source [198] (E-19). Gas phase nucleation (gas-condensation) has also been used to form neutral and ionized ultrafine particles ("nanoparticles") using a sputtering source and a plasma condensation chamber [199].

In 1965 Smith and Turner described the use of a ruby laser to vaporize (flash evaporate) material from a surface and deposit a film [200]. This process is sometimes called laser ablation and the deposition process, laser ablation deposition (LAD) or pulsed laser deposition (PLD) [201]. PLD and reactive PLD have found application in the deposition of complex materials such as superconductive [202] and ferroelectric thin films.

Epitaxy ("oriented overgrowth"), where the crystalline orientation of the deposited film is influenced by the crystalline orientation of the substrate material, has been recognized since the 1920s and was reviewed by Pashley in 1956 [203]. Molecular beam epitaxy (MBE) is an advanced, sophisticated vacuum deposition process that uses beams of atoms or molecules of the material to be deposited to form large-area single crystal films. MBE was first proposed by Günther in 1958 [204] but the first successful deposition had to await the development of ultrahigh vacuum technology. In 1968 Davey and Pankey successfully grew epitaxial GaAs by the MBE process [205]. The modern use of MBE in semiconductor device fabrication began with Cho and Arthur in 1975 with the growth of III-V semiconductor materials [206]. The use of organometallic precursor vapors as a source of the depositing material in epitaxial growth is called "organometallic vapor phase epitaxy" (OMVPE) [207]. Transmission electron microscopy (TEM) (Max Knott and Ernst Ruska—1931) and electron diffraction techniques allow the determination of crystalline perfection and crystalline defects. High-resolution TEM was perfected in 1939 by Bodo von Borries and Ernst Ruska (Siemens Super Microscope). TEM is one part of the analysis technique called analytical electron microscopy (AEM).

Before the end of WWII, the thickness of deposited optical coatings was determined by visually observing the transmittance or reflectance during deposition. Around 1945 optical instrumentation was developed for monitoring the thickness during deposition [208]. In 1959 Steckelmacher, Parisot, Holland, and Putner described a practical optical monitor for use in controlling the film thicknesses in multilayer interference coatings [209]. In the late 1950s quartz crystal monitors (QCMs) began to be developed for determining the mass of deposited material *in situ* [210]. After WWII the development of laser technology, particularly high-energy lasers, required very high-quality optical and reflecting coatings [211].

Thickness uniformity is often a concern in vacuum coating. Thickness uniformity is often determined by the fixture configuration and movement. A concern in reactive deposition is the availability, uniformity, and degree of "activation" of the reactive species. Therefore the geometry of the manifold used for introducing the reactive gases is an important design. Fixture configuration and movement can also be used to enhance reaction uniformity.

In much of the early work obtaining a uniform coating over a large stationary area was done using multiple sources [131]. Controlled thickness distributions using moving shutters were also done [131 p. 180]. Later moving shutters were used to get improved thickness uniformity over large areas from a point source [212]. Shaped evaporation sources were also developed to improve thickness uniformity.

Electron beam polymerization was observed in the early electron microscopes when hydrocarbon pump oil vapors were polymerized on the specimen by the electron beam. In 1958 Buck and Shoulders proposed the use of electron beam polymerized siloxane vapors as a resist in forming miniature printed circuits [213]. Electron beams and ultraviolet (UV) radiation are used to "cure" vapor-deposited organic and inorganic fluid films in vacuum. In the polymer-multi-layer (PML) process the degassed monomer is sprayed as a fine mist on the moving part, usually in a web coating arrangement [214].

In the "parylene process" a vapor of organic dimers is polymerized in a hot zone then deposited on a cold surface [215]. The process was initially reported by M.M. Swarc in 1947. The presently used process was developed by W.F. Gorman at Union Carbide beginning in 1952 and was commercialized in 1965 using di-p-xylyene (DXP) at pressures near 1 Torr.

Arc Vapor Deposition

(General References [15, 216-218])

Discussion of arcs and arc vapor deposition is somewhat difficult due to the varying definitions of an arc. These definitions vary from "a low-voltage, high-current discharge between electrodes," to "a discharge in a gas or vapor that has a voltage drop at the cathode of the order of the minimum ionizing or minimum exciting potential of the gas or vapor" (K.T. Compton) [216], to "a self-sustained discharge capable of supporting large currents by providing its own mechanism of electron emission from the negative electrode" (J.M. Lafferty) [15]. The history of the study of arcs, sparks (low total-power arcs), and "flashovers" (short-duration arcs and sparks over the surface of an insulator) is quite old, with studies of the chemistry in spark plasmas predating the invention of the voltaic battery. With the invention of the voltaic battery (Volta, 1800), high-current, low-voltage electric power was available and the study of chemistry in arcs ensued [24]. In 1839 Robert Hare described the first arc furnace that was used for melting materials in controlled atmospheres [219], and in 1903 Werner von Bolton used an arc in vacuum to melt materials [219].

Michael Faraday might be credited with vapor depositing the first material that was studied for its properties (optical) in 1857 [220]. The films were probably deposited by arc (spark) vaporization ("Gold wire deflagrated by explosions of a Lyden battery—."), and "exploding wire" ("When gold is deflagrated by the voltaic battery near glass—."). Faraday deposited the films in different gaseous environments (oxygen, hydrogen) that were probably produced by vacuum pumping and flushing. Thus he probably did some depositions in a vacuum (subatmospheric) environment. The work was a minor part of Faraday's studies on the optical properties of fine particles ("colloidal") of gold, mostly prepared from chemical solutions. The technique was not pursued but was referred to by A.W. Wright in his 1877 paper on "electrical deposition" [37].

In non-self-sustaining arcs, the electron emission from the cathode may be from thermal heating (thermoelectron emission) [221] or from hollow cathode emission [222]. In arc vapor deposition for PVD, vaporization can either be from a molten (usually) anode ("anodic arc") or from a molten spot on a solid (usually) cathode ("cathodic arc"). An advantage of both types of arcs is the high percentage of ionization of the material vaporized from the electrodes, as well as ionization of the gases that may be present. The ions of the vaporized material ("film-ions" or "self-ions") may be accelerated to high kinetic energies. This notable property is the primary reason that the author has treated arc vapor deposition separately from thermal evaporation.

In 1884 Edison filed a patent entitled "Art of Plating One Material on Another" that included arc vaporization as a deposition technique [40]. Figure 16 shows the figure from his patent that was related to arc vaporization. This aspect of the patent was challenged based on Wright's work, as was discussed previously. Edison was finally granted the patent in 1894 after modifying his claim to call his arc a "continuous arc" and that of Wright a "pulsed arc" and a "laboratory curiosity." Edison attempted to use arc vaporization to deposit the "seed" layers on his wax phonograph masters. In a 1902 patent entitled "Process of Coating Phonograph Records," he stated, "I find that in practice that the employment of an electric arc for vaporizing the metal, as suggested in my patent, to be open to the objection of being slow,

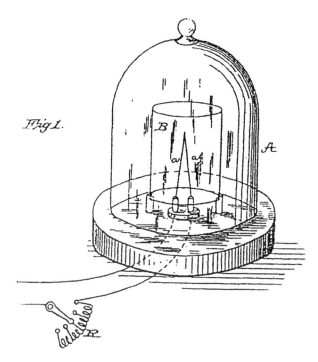

Figure 16. Edison's arc vapor deposition apparatus as shown in his 1894 (filed 1884) patent [40].

and unless the process is carried out with great care the deposit is not entirely uniform, while there is danger of injuring the very delicate phonographic record surface, particularly from the heat of the arc." Instead he used a configuration that was obviously sputtering, as shown in Figure 7 [42]. Several vacuum arc vapor sources (as well as others) were developed during WWII to form ion beams for use in the magnetic isotope separation program [223]. These ion beams formed deposits but they were not studied for their properties. Arc vapor deposition activities then died down until well after WWII.

In 1954 Bradley deposited carbon films using a pointed rod-type carbon arc vapor source [224]. In 1962 Lucas, Vail, Stewart, and Owens deposited refractory metal films [225] by arc vapor deposition. In 1965 Kikushi, Nagakura, Ohmura, and Oketani investigated the properties of a number of metal films deposited from a vacuum arc [226]. In 1965 Wolter investigated the properties of Ag, Cu, and Cr films deposited by "anodic" arc vapor deposition from a molten metal using a heated electron-emitting filament as the cathode [227]. In 1972 Morley and Smith used a hollow cathode electron source for rapidly evaporating material for film deposition [177, 228]. Ehrich continued to develop the molten anodic arc deposition system [229].

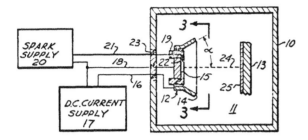

Figure 17. Cathodic arc vaporization source as shown in Snaper's patent (1974) [230].

In the "cathodic arc" discharge, vaporiz. ion originates from a spot on the surface. The spot may be stationary ("attached"), but it usually moves over the surface naturally or under the influence of a magnetic field. In the mid-1970s workers in the Soviet Union, while working on plasma sources, developed sources for arc vaporization and film deposition using arcs moving over a surface. Some of their work involved reactive deposition of nitrides (E-20). Some of the early patents on arc vaporization sources were by Snaper (1974—magnetic field coils) [230], Sablev et al (1974—no magnetic field) [231], and Mularie (1984) [232]. Figure 17 shows a figure from the Snaper patent. Generally arc sources use DC, though pulsed sources have been developed [233].

Early cathodic arc sources used "random" or "steered" arcs on planar surfaces. The arcs are initiated by a high-voltage "trigger arc" or by making and breaking electrical contact to the surface. Recently laser pulses have been used to trigger the arcs [234]. In 1991 Vergason patented a cathodic arc source where vaporization was from arcs that traveled from one end of a long rod to the other as the negative contact was changed from one end of the rod to the other, and a "trigger arc" initiated the main arc at the end opposite from the electrical contact. In traversing the length of the rod, the arcs give uniform erosion over the surface of the rod [235]. Figure 18 shows figures from that patent. In 1993 Welty added a low-turn, current-carrying coaxial coil around the rod to provide an additional axial magnetic field that aided in spiraling the arc around the cathode rod and controlled the arc travel time [236].

An advantage of arc vaporization is that a large fraction of the vaporized material is positively ionized [237] and these "film ions" can be accelerated to the substrate surface if there is a negative potential on the substrate (see ion plating section). In gaseous arcs the dissociation, ionization, and excitation

of the reactive gas ("activation") can be important in reactive deposition processes.

A disadvantage of cathodic arc physical vapor deposition (CAPVD) is the generation of molten "macroparticles" during arcing. These "macros" create bumps and pinholes in the film. One method of eliminating the macros is to use a "filtered arc source" where the plasma of "film ions," gaseous ions, and electrons is deflected out of the line-of-sight of the source. Most of the macros and uncharged film atoms are deposited on the walls. This approach reduces the deposition rate on the substrate significantly. The most common design uses a quarter-torus plasma duct [238]. Beginning in the 1990s the Royal Australian Mint used filtered arc deposition to deposit TiN hard coatings on their coining dies [239].

Arc vapor sources can provide ions of the depositing material to sputter a target of the same material [240]. This technique eliminates the "macros." A magnetic field arrangement to increase the electron path length in front of the arc cathode may be used to increase the plasma density and heat "macros" to increase the evaporation from the macros after

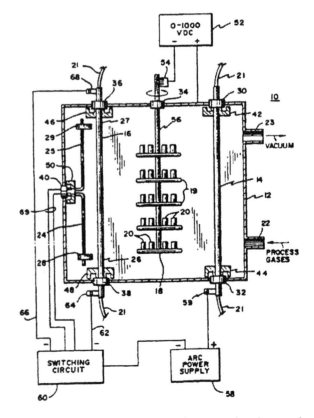

Figure 18. Figure from Vergason's patent showing a rod-type cathodic arc vaporization source (1990) [235].

they are ejected [241]. A modified tungsten-inert-gas (TIG) plasma arc welding source has been used to provide a plasma for melting and evaporating a material in vacuum [242].

A magnetron source that used magnetic fields that emerge and re-enter a planar mercury surface ("magnetic tunnel," described in the sputter deposition section) was developed by Kesaev and Pashkova in 1959 for steering arcs [243]. Steered arc vaporization can be done with the same configuration as planar magnetron sputtering [244]. Sputtering and arc vaporization can be combined into the same target by changing the magnetic field configuration [106]. This allows arc vaporization to be used for depositing the "adhesion layer," then the film thickness is built up using sputtering where there are no macros (ArcBond™ process). Arc vaporization as a vapor source can be used with a continuous or pulsed bias [245] to extract ions from a plasma to bombard the growing film (ion plating).

The use of a capacitor discharge through a thin wire to vaporize and "explode" the wire can be considered an arc discharge. This vaporization technique and the films thus formed has had limited study [246].

Chemical Vapor Deposition (CVD)

(General References [1, 247-249])

Chemistry in plasmas occurs naturally with arcing phenomena such as lightning producing the chemical ozone. Von Gurkie, who made the first frictional electricity machine using a rotating sulfur ball, must have created (and smelled) hydrogen sulfide when it sparked in the presence of moisture. The deposition of carbon from spark discharges in hydrocarbon gases was reported by Henry in 1798. In 1802 Sir Humphry Davy used the new voltaic cells to study arc discharges and the generation of compounds in the arc plasmas.

In the 1880s and the 1890s, chemical vapor deposition (CVD) began to be developed in earnest with the goal of strengthening the carbon filaments in the first light bulbs. In 1880 Sawyer and Man deposited pyrolytic carbon by pyrolysis [250] and in 1891 Mond deposited nickel from nickel carbonyl [251]. The hydrogen reduction of metal chlorides on heated surfaces to form metal films was described by Aylesworth in 1896 [252]. Early chemical vapor

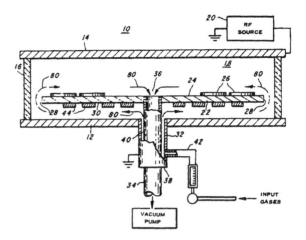

Figure 19. "Reinberg" parallel-plate PECVD reactor (1973) [269].

deposition was performed at atmospheric pressure following vacuum pumping and flushing to rid the system of unwanted gaseous species. The concentration of chemical vapor precursors and reactants in CVD is controlled by diluents and carrier gases such as argon and nitrogen. Titanium carbide coatings were developed and commercialized on tools in the mid-1930s by Metallgesellschaft AG. Modern development of thermally driven atmospheric pressure CVD (APCVD) is considered to have begun with Lander and Germer in 1948 [253]. In 1961 Theurer demonstrated the deposition of single-crystal (epitaxial) silicon by CVD [254]. CVD reactors are either "cold-wall reactors" where only the substrate and its holder are heated or "hot-wall reactors" where a general volume is heated and a "batch" of substrates is coated. A laser was first used for localized heating for CVD in 1973 [255] and presently a laser is used for heating a moving optical glass fiber for CVD deposition.

APCVD is performed at atmospheric pressure and rather high temperatures. Low-pressure CVD (LPCVD) (or sub-atmospheric CVD—SACVD) is performed at pressures of 0.25 to 2.0 Torr in a reaction-rate limited mode [256]. In general LPCVD provides better coverage and less particulate contamination than APCVD. LPCVD (SACVD) is used in the semiconductor industry for depositing doped silicon, compound semiconductors, tungsten metallization, and passivation layers [257]. Pulsed low-pressure CVD, using pulses of reactant into a low-pressure environment, has been used to deposit thin films [258]. A type of pulsed CVD is called "atomic

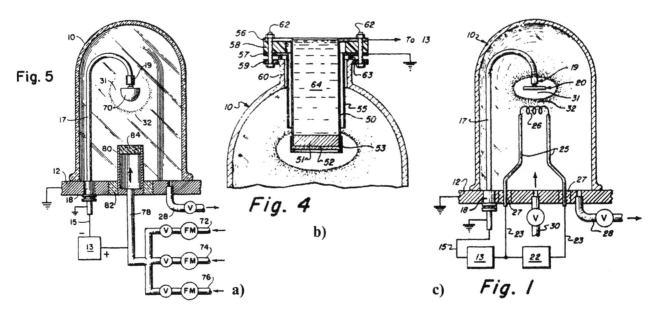

Figure 20. Figures from Mattox's "ion plating" 1967 patent: a) use of chemical vapor precursors in the plasma, b) use of a "cold finger" fixture to keep the bulk of the substrate very cold during deposition, and c) simple un-cooled, un-heated substrate fixture [272].

layer deposition" (ALD) where a single layer (or less) of atoms is deposited on a surface by pulsed-pressure CVD processes and the thickness of the deposited film is determined by the number of pressure pulses (AL-CVD). This technique has advantages in coating high-aspect-ratio features on the surface [259].

Early vacuum deposition from plasmas containing chemical vapor precursors was termed "glow discharge deposition." Later the process was called "plasma deposition" [248, 260], and later still, "plasma-enhanced CVD" (PECVD) or "plasma-assisted CVD" (PACVD). In 1911 von Bolton reported the deposition of seed crystals of "diamond" from "decomposition of illuminating gas (C_2H_2) in the presence of Hg vapors" [261]. The deposition of inert and glassy carbonaceous films on the walls of gas discharge tubes containing hydrocarbon vapors and other carbon-containing vapors was noted in 1934 [262]. In 1953 Schmellenmeier studied plasma-deposited carbon films and discovered the presence of diamond-like structures (diamond-like carbon—DLC) [263]. DLC films can be produced by several techniques [264], including pyrolysis [265], plasma deposition [266], and carbon-ion ("i-C") beam deposition.

In 1965 Sterling and Swann discussed the importance of plasmas in decomposition of chemical vapor precursors [267]. When depositing species from a plasma, the molecules are often incompletely dissociated, thus forming a spectrum of molecules and molecular fragments, both neutral and ionized. Combination of fragments increases the number of species in the plasma. For example, in the PECVD of silicon from silane (SiH_4), the formation of Si_2H_6 is thought to play an important role as it is more readily adsorbed on a surface than is silane [267a]. Ozone (O_3) is reported to be more readily adsorbed on oxide surfaces than is O_2 [267b]. Plasma decomposition and vapor phase nucleation can also produce ultrafine particles [196, 268].

In 1971 Reinberg developed an RF-driven, parallel-plate, PECVD reactor that allowed CVD of materials at a lower temperature than was possible by using thermal processes alone [269]. The "Reinberg reactor," Figure 19, is generally operated at pressures greater than 0.5 Torr and deposits materials such as glass films (phospho-silicate-glass-PSG) and Si_3N_4 (from silane and nitrogen) for semiconductor encapsulation. Plasma-deposited films are also used for depositing optical coatings [270].

Film deposition by PECVD at very low pressures (10–50 mTorr) (LP-PECVD) allows ions from the plasma to be accelerated to high energies. LP-PECVD onto a negatively biased substrate under high-energy particle bombardment was described in the "ion plating" patent of Mattox in 1967 [271]. Figure 20a shows figure 5 from that patent. Culbertson and Mattox used this "chemical ion plating" LP-PECVD process for depositing tungsten from WCl_4 in 1966 [272]. Culbertson used the process to deposit carbon and metal carbide coatings on electron tube grids to provide a low secondary emission coefficient in 1971 [273]. Culbertson's arrangement for depositing titanium carbide films from chemical vapor precursors is shown in Figure 21. He also described using a thermal evaporation filament for the metal vaporization. This was the first application of hybrid PVD/LP-PECVD. Later this process became important in hybrid deposition of hard coatings and decorative/wear coatings, such as metal carbonitrides (e.g., Ti-Al-C-N), using a combination of reactive PVD and LP-PECVD (see discussion of ion plating). In 1975 Spear and Le Comber showed that the valency of plasma-deposited amorphous silicon could be controlled [274]. This material became the basis of a-Si solar cells (Carlson and Wronski, 1976) [275]. In 1979 LeComber, Spear, and Ghaith reported the development of an amorphous field-effect transistor using amorphous-hydrogenated silicon (a-Si:H) prepared by PECVD of silane (SiH_4) [276].

In 1983 Môri and Namba used methane in a plasma-type ion source to deposit DLC films under bombardment conditions at very low pressures [277]. In the 1990s pulsed DC and bipolar pulsed-power PECVD began to be developed [278] and is sometimes called plasma impulse CVD (PICVD). Chemical vapor precursors, at low pressures, are also used in "chemical beam epitaxy" (CBE) [279]. PECVD shares many of the same concerns as "plasma etching," "plasma stripping," and "reactive plasma cleaning." One of these concerns is the pumping of "dirty" and reactive gases, vapors, and particulates through the vacuum pumping system.

Probably the first metal ion beam-type source using the disassociation of a chemical vapor in a plasma was the "Calutron" source developed in WWII [280] (E-21). The Calutron produced ions that were accelerated to 35 kV and magnetically deflected to separate isotopic masses. In 1983 Shanfield and Wolfson formed vacuum-deposited compound films (BN) from a chemical vapor precursor of borazine ($B_3N_4H_6$) injected into a Kaufman-type plasma source [281]. In 2002 Madocks described a magnetically "pinched" (focused) plasma configuration that may be useful for large-area plasma deposition [282].

Plasma polymerization (PP) was first noted by de Wilde [283] and also Thenard [284] in 1874. The

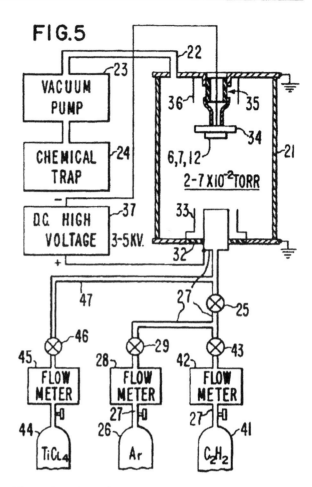

Figure 21. Figure from Culbertson's 1971 patent showing arrangement for depositing titanium carbide films [273].

phenomena was not put into practical use until the 1960s when plasma polymerization was developed to apply thin, pinhole-free polymer coatings on food containers for corrosion protection [285]. Partially oxidized, plasma-polymerized coatings are of interest as permeation barriers and have been plasma-deposited from hexamethyldisiloxane (PP-HMDSO) and hexamethyldisilazane (PP-HMDSN) [286]. The degree of oxidation is determined by controlling the partial pressure of oxygen in the discharge. This type of coating is used as an external vapor barrier on the external surface of three-dimensional plastic containers. Plasma polymerization, along with PVD-type vaporization of solids, is used to deposit mixed metal-polymer "composite" films [287]. Ion flux and electron temperature are important parameters in plasma polymerization at low pressures [288].

Ion Plating

(General References [271-273, 289-295])

In 1963 Mattox described the use of a negative substrate potential on a substrate to obtain ion bombardment from a plasma before and during film deposition [271, 296] (E-22, E-22a). Bombardment during deposition can be used to modify the properties of the deposited material. Figure 20a shows a figure from the patent that uses LP-CVD. Figure 20c shows the use of a resistively heated evaporation filament. Ion bombardment from the plasma sputter-cleans the substrate surface [297], and by continuing the sputtering while beginning the film deposition, the film-substrate interface is maintained contamination-free.

Initially ion plating used a thermal vaporization source or a chemical vapor precursor or a combination of the two sources. Later, when high-rate sputter vaporization sources were used as a source of deposited material, the process was called "sputter ion plating" by some authors [298] and "bias sputtering" by others. In the ion plating process deposited material may be "back-sputtered" if the bombarding energies are above some "sputtering threshold" energy. Generally this threshold energy is 50 to 100 eV depending on the depositing material and mass of the bombarding species. Bombardment below this energy can modify film properties without having any "back-sputtering" [299]. Generally at low bias potentials high ion fluxes are used to modify the properties of the coating. When depositing compound or alloy materials, preferential back-sputtering may affect the composition of the deposit.

To some, ion plating represents a defined deposition process with various vapor sources; to others, ion plating is a deposition parameter for deposition processes defined by the vapor source. When using a chemical vapor precursor in a plasma, some authors refer to the process as "chemical ion plating," and it is a low-pressure version of plasma-enhanced chemical vapor deposition (PECVD) (see CVD section). The term "ion plating" was first used in 1964 and rationalized by Mattox in 1968 [300]. Some authors prefer the term "ion assisted deposition" (IAD) or "ion vapor deposition" (IVD), or ionized physical vapor deposition (iPVD) [301], and it may also be considered an "energetic condensation" process [302].

Ion bombardment of the substrate and the growing film adds thermal energy to the surface region without having to heat the bulk of the material. This allows a sharp thermal gradient to be established between the surface region and the bulk of the material [303]. Figure 20b show the use of a liquid-nitrogen-cooled "cold finger" to keep the bulk of the substrate cold while the surface is bombarded. Figure 20c shows a non-heated, non-cooled substrate holder configuration for ion plating. The ion bombardment also causes atomic rearrangement in the near-surface region during deposition, which increases the film density and intrinsic compressive film stress [304]. The term "atomic peening" for this densification process was first used by Blackman in 1971 [305]. Ion bombardment also increases the chemical reactivity between the depositing film material and the co-deposited or adsorbed material [306]. The presence of the plasma "activates" reactive species in the gaseous deposition environment and makes them more chemically reactive with the deposited material ("reactive ion plating").

Mattox also demonstrated that the ion plating process provides good surface coverage over three-dimensional surfaces from a point vaporization source. This is due to scattering in the gaseous environment when the pressure is in the range of 10 to 40 mTorr. The problem of "sooting" [196] was not encountered because the vapor phase condensed particles become negatively charged and are repelled from the negative substrate. Soot deposit on the wall, however, is a problem for system maintenance when high-rate vaporization is used. Attempts to use scattering in the gas phase to improve coverage without a negative bias and bombardment were not very successful [307].

A great deal of effort was expended to find "prior art" for the ion plating technology, even though the Mattox patent was in the public domain from its inception since he worked under contract for the United States Atomic Energy Commission (USAEC) (E-11). One prior patent by Berghaus was located. That patent described a similar process that involved coating articles with a negative bias in a glow discharge, as shown in Figure 22 [308]. Berghaus in his 1938 and 1942 patents stated that the coating adhesion was improved and that the coating was densified by the method. No technical papers were found on his process and the technique was not mentioned in L. Holland's book *Vacuum Deposition of Thin Films*

(1957), though Holland referenced other work of Berghaus.

In 1962 F. Wehner patented a process using a triode-sputtering configuration for deposition of epitaxial films [69]. Figure 9 shows his apparatus, which uses a mercury plasma. A paper published in 1962 discussed periodic bombardment of the depositing film using an asymmetric AC potential between the sputtering source and the substrate to improve the properties of superconductor film material by periodic ion bombardment during deposition. Frerichs called this technique "protective sputtering" [309].

In 1965 Maisel and Schaible used a negative substrate bias during sputter deposition ("bias sputter deposition") to improve the purity of sputter-deposited chromium films [310]. They found that oxygen was removed from the depositing film, forming a purer metal film. In 1965 Glang, Holmwood, and Furois studied the stress in films formed with concurrent ion bombardment [311]. In 1968 Vossen used an RF substrate bias while sputter-depositing aluminum metallization on silicon devices to improve the "throwing power" into surface features [312]. In 1970 Maissel, Jones, and Standley demonstrated that an RF bias on a sputter-deposited ceramic film would densify the deposit [313]. Reactive sputter deposition under ion bombardment conditions began in 1969 with the work of Krikorian and that of Pompei [314].

The ion plating technology was immediately applied to production applications, including the coating of uranium fuel elements for pulsed nuclear reactors [315], tribological coatings for the space program [316], and aluminum coating of fasteners for the aerospace industry [317]. The latter process was known as the "Ivadizing" process using Ivadizer™ equipment, and ultimately the "ion vapor deposition" (IVD) process (E-23). The IVD "specs" are "called out" in many military procurement orders [318]. In 1973 Bell and Thompson reported using ion plating as a "strike coat" for electroplating in order to get good adhesion [319]. In 1984 Panitz and Sharp reported the use of PVD deposited films followed by barrier anodization to form aluminum thin film capacitors [319a]. In 1968 Mattox and Rebarchik reported the use of a biased rotating "cage" with a high-transmission grid to contain loose parts such as nuts and screws (320). The cage avoided the need to make electrical contact to individual parts and is analogous to "barrel plating"

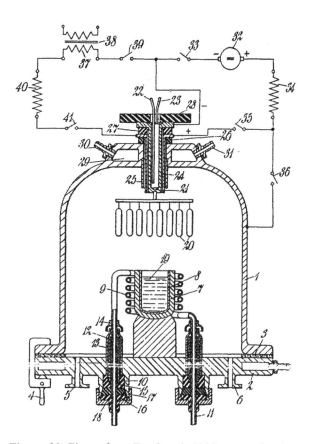

Figure 22. Figure from Berghaus's 1938 patent showing an "ion plating" configuration [308]. This work was overlooked for 25 years.

in electroplating. A vibrating table can be used for coating balls in the same manner. The use of a high-transmission grid structure in front of an electrically insulating surface allows ion bombardment of the surface and any dielectric film growing on the surface (321).

In 1975 Schiller, Heisig, and Goedicke published work on thermal evaporation and sputter deposition in which a small amount of metal was periodically deposited and then exposed to bombardment by energetic ions. A coating was obtained by repetitive deposition and bombardment [322]. This process was called "Alternating Ion Plating." A similar technique was employed by Lefebvre, Seeser, Seddon, Scobey, and Manley in 1994 in a patent for periodic deposition and reactive ion bombardment on a rotating cylindrical ("drum coater") substrate fixture (OCLI's MetaMode™) to form a coating of a compound material by reactive deposition [323]. A further development is the use of a microwave injector system to "activate" the oxygen that is injected to react with the periodically deposited film material on the rotating drum and to increase the plasma density above the sputtering target (DSI MicroDyn™ equipment) [323a]. This allows slower drum rotation and a thicker film to be deposited on each pass.

Electron beam (e-beam) evaporation was developed in the mid-1960s, but initially it was not used for ion plating because the high negative voltage on the filament prevented the source from being used in a plasma where positive ions were accelerated to the filament and rapidly eroded the filament. In 1971 Chambers and Carmichael [176] avoided that problem by having the electron beam pass through a small hole in a thin sheet in a section of a plate that separated the deposition chamber and the chamber where the filament was located. This allowed a plasma to be formed in the deposition chamber while the filament chamber was kept under a good vacuum. The plasma allowed the ion bombardment of material being thermally evaporated using an e-beam as well as "activation" of a reactive gas. This e-beam evaporation technique could then be used for the deposition of refractory metals and alloys [324]. A magnetic field can be used to guide electrons to impinge on an electrically isolated surface in order to create a "self-bias" that then attracts ions for concurrent bombardment during deposition [94, 325]. This allows ion plating to be performed on an electrically insulating surface without having to use an RF bias.

The use of a negative potential on a substrate immersed in a plasma is called "plasma-based ion plating." From its inception it was realized that only a small fraction of the evaporated or decomposed chemical vapor precursors was ionized ("film ions"). Several investigators attempted to quantify the number of film ions formed with varying techniques [326]. In some investigations post-vaporization ionization of the vaporized material was employed [327]. Post-vaporization ionization and acceleration is also used in doping of films grown by MBE [328].

With the advent of arc vaporization (which produces a high fraction of vaporized species) as a source of vapor for ion plating, the effects of "self-ion" bombardment became increasingly important [329]. This bombardment has the advantage that the mass of the bombarding species is the same as the depositing species, thus maximizing the energy and momentum transfer. Both cathodic arc and anodic arc vaporization have been used for ion plating with the Russian BULAT-type being one of the first such

cathodic arc deposition systems [38]. In 1991 Olbrich, Fessmann, Kampschulte and Ebberink reported using a pulsed bias to reduce the heat load on the substrate during deposition [329a].

When sputtering at low pressures the reflected high-energy neutrals are not "thermalized" and bombard the depositing film with the same effect as ion bombardment. By changing the gas pressure, the degree of thermalization can be controlled. "Pressure pulsing" has been used to modify the properties of the depositing material in the same manner as ion bombardment [124]. Energetic particle bombardment of the surface or depositing film can result in incorporation of high concentrations of the bombarding gas into the surface or film [330].

If the material is deposited in a good vacuum, the process may be called "vacuum ion plating" [331]. Vacuum ion plating can be done using condensing vapor bombarded by a separate ion source that may use an inert or reactive gas [332]. This is usually called ion beam assisted deposition (IBAD). The IBAD process has been especially effective in depositing dense coatings for optical coating applications. Bombardment can be done using a beam of condensable "film-ions" from an ion source. In 1969 Aisenberg and Chabot discussed ion beam deposition of elements and compounds using a plasma source where ions of the electrode material (carbon and doped silicon) were extracted into the deposition chamber [333]. When using a pure beam of "film ions" there is no possibility of incorporation of a foreign species in the deposited film. In 1973 Aisenberg used energetic carbon ion (i-C) beams to deposit diamond-like carbon (DLC) material in good vacuum under bombardment conditions [334] (E-24). Figure 23 shows a source from his 1976 patent. The carbon film ions are used to deposit DLC films on plastic lenses and blade edges. Film ions can also be obtained from field emission sources [335]. At high bombardment fluxes and/or energies, "back-sputtering" by gaseous ions [336] or "self-sputtering" by "film ions" [337] can be an important effect. To have appreciable back-sputtering, a low flux of high-energy ions or a high flux of low-energy ions is needed since the "sputtering yield" is a function of the mass and the energy of the bombarding particles.

The source of condensable material for ion plating can be a chemical vapor precursor, such as that used in chemical vapor deposition (CVD). Ion-plated

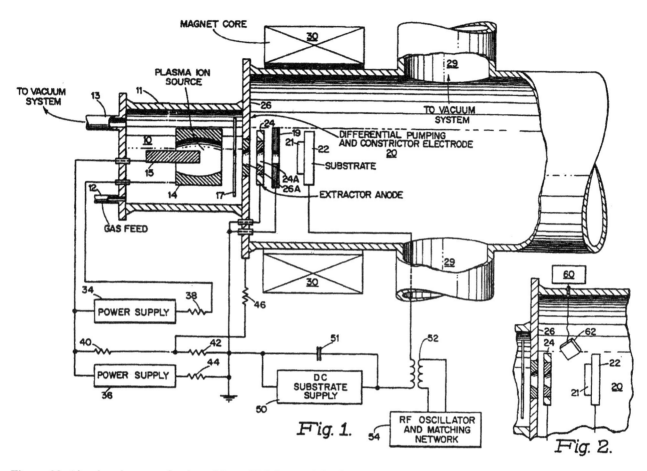

Figure 23. Aisenberg's patent for deposition of DLC material using an ion source from which high-energy "film ions" were extracted (1975) [333].

tungsten was deposited from WCl_4 in 1966 [272]. Ion-plated carbon (C) and metal (Ti, Cr, Zr, Si) carbide coatings from chemical vapor precursors (e.g., $TiCl_4$ and C_2H_2) were deposited on electron tube grids to decrease secondary electron emission [273]. The carbon coatings were probably what were later called diamond-like-carbon (DLC) coatings. This application was the first use of a combination of PECVD and PVD processes (hybrid process) to deposit metal carbide coatings by ion plating, which is now a common process. In 1974 Mattox discussed the use of reactive ion plating for the deposition of compound films for hard coatings [338].

Periodic bombardment of a substrate surface with very high-energy ion bombardment (>10 keV) is called "plasma immersion ion implantation" (PIII

or PI³) [339] and is similar to a beam-type shallow ion implantation, which was developed in the early 1970s for the semiconductor device industry, except that it is non-directional. The use of pulsed power prevents the discharge from becoming an arc. If PIII is performed during metal deposition, it is called "plasma immersion ion implantation deposition" (PIIID) [340]. Pulsed plasmas in a high-voltage co-axial plasma gun have been used to erode the electrodes and provide the vapor to be deposited. Periodic discharges were obtained by pulsing gas through the tube. For example, Ti(CN) can be deposited using titanium and carbon electrodes and a nitrogen working gas [341].

Surface Preparation

(General References [342, 343])

Surface preparation is an essential step in all vacuum coating processing and is critical to the nucleation and interface formation steps of the coating process. These "prepared" surfaces, in turn, determine the adhesion of the coating, the surface coverage, and—in many cases—the structure and the properties of the resulting deposit. Surface preparation includes not only cleaning but in some cases modification of the chemical, morphological, or mechanical properties of the surface or the near-surface region [344]. These changes can affect the nucleation of the depositing atoms on the surface and the growth of the coating. Cleaning comprises both that done external to the deposition chamber (external cleaning) and that done in the deposition chamber (in situ cleaning).

External cleaning for vacuum coating is generally similar to that used in other manufacturing processes of the type called "critical cleaning [342]" or "precision cleaning." Precision cleaning is sometimes defined as "cleaning a surface that already looks clean." A few external cleaning techniques have been developed specifically for vacuum coating or were initially used for vacuum coating. These include ultraviolet-ozone (UV/O$_3$) oxidative cleaning and sputter (physical) cleaning.

UV/O$_3$ cleaning was initially developed by Sowell, Cuthrell, Bland, and Mattox in 1974 to clean very fragile quartz-crystal structures that were contaminated with Carnauba wax during grinding and polishing [345]. The cleaning technique was found to be very effective for removing contaminants that formed volatile oxidation products, such as hydrocarbons, from oxidation-resistant surfaces, such as glass, or from surfaces where a coherent oxide was desirable, such as silicon. The UV/O$_3$ environment was also found to create a hydrocarbon-free storage environment for storing cleaned surfaces prior to coating (E-25).

An important aspect of cleaning is to avoid contamination of the surface with particulates. Particulates are a primary cause of pinholes in films deposited on smooth surfaces. In 1960 Willis Whitfield (Sandia Corp., later Sandia National Laboratories) developed the laminar flow "clean bench" (E-26) that uses a fiber filter to remove fine particulates from the air. The mechanical fiber filter is called a HEPA (High Efficiency Particulate Air) filter and was initially developed for use in filtering the air in mine-safety appliances. Initially clean benches and clean rooms were developed to remove radioactive particles and provide clean areas for assembly of contaminant-sensitive components. This was an improvement on the "white rooms" used during WWII.

"Cathodic etching" (preferential sputtering) was probably first used for preparing the surface of metallographic samples for microscopic observation. Preferential sputtering of different crystallographic surfaces delineates the grain structure on highly polished surfaces. The etching also brings out inclusions in the material. This sputtering process cleans the surface, but that is not the purpose of the sputtering in preparing metallographic samples. "Sputter texturing" has been used for creating very micro-rough surfaces by simultaneously contaminating a surface with carbon while sputter etching. The carbon forms islands on the surface that protect small regions of the surface from sputter etching [346]. "Sputter depth profiling" as an analytical technique was developed for use with Auger Electron Spectroscopy in the 1960s.

In 1955 Farnsworth, Schlier, George, and Burger reported using sputter cleaning in an ultra-high-vacuum system to prepare ultra-clean surfaces for low-energy electron-diffraction (LEED) studies [297]. Sputter cleaning became an integral part of the ion-plating process [289-292] and is now routinely used in the deposition of hard coatings on tools. In 1976 Schiller, Heisig, and Steinfelder reported sputter cleaning of a strip by passing it through the "racetrack" of a magnetron sputtering source [347]. Sputter cleaning has some potential problems such as overheating, gas incorporation in the surface region, bombardment (radiation) damage in the surface region, and the roughening of the surface, particularly if "over done." It is important to have a "clean" plasma in order to not continually recontaminate the surface during sputter cleaning [348]. Gas incorporation during sputter cleaning can result in adhesion problems between film and substrate [349]. Deposition of the sputtered material on the vapor source such as a filament evaporator [315] or sputtering target can result in redeposition of the contaminant on the substrate when the deposition process is begun. Redeposition of sputtered material on the substrate can also give problems, especially at high sputtering pressures [350].

Sputtering of the surface of a compound or alloy material can result in the surface composition being changed. Often the species with the least mass [351] or the highest vapor pressure [352] is the one preferentially sputtered from the surface. Bombarding of a carbide surface with hydrogen can result is the loss of carbon ("decarburized") [353]. Extensive sputtering of a surface can be used to form desirable features or to thin a specimen for transmission electron microscopy [354]. This process is called "ion milling."

Reactive cleaning uses a reactive gas (H, Cl, F) or vapor (I) [354] to form a volatile compound of the contaminant and/or surface material. Reaction may be promoted by heating—for example, "air firing," or with the use of a plasma in a vacuum, "reactive plasma cleaning." Reactive plasma cleaning was the first *in situ* cleaning process and was initially used to clean glass surfaces for optical or reflective coatings. This process, using an air (oxygen-containing) discharge, was used by Strong in 1935 before depositing reflective coatings on glass [355]. Strong used this technique for cleaning glass surfaces for astronomical telescopes [149] (E-27). Most optical coating equipment, even if it only uses thermal evaporation for deposition, has a high-voltage "glow bar" that allows the formation of a plasma for plasma cleaning of surfaces. In 1977 Mattox and Kominiak reported reactive plasma cleaning of metal surfaces with plasmas containing chlorine [356]. A carbon deposit was left on the surface when using a chlorine-containing chemical such as CCl_4, so HCl was found to be a better choice. Hydrogen can also be used for plasma cleaning to remove hydrocarbons and other materials that form volatile hydrides [357]. Sputtering with a reactive gas or vapor such that a volatile compound is formed on the surface and then thermally vaporized is sometimes called "chemical sputtering" [358] or "reactive ion etching."

A surface in contact with a plasma acquires a negative voltage (several volts) with respect to the plasma. This sheath potential accelerates positive ions to the surface. When an ion contacts the surface it acquires electrons and releases its energy of

ionization (5 to 20 eV). This low-energy bombardment and heating desorb adsorbed contaminants such as water vapor. This type of plasma cleaning is called "ion scrubbing" [343].

Plasmas are also used to "activate" the surface of polymer substrates either by generating free radicals of surface species or grafting reactive species, such as oxygen or nitrogen, to the surface ("functionalization"). This activation can be used to increase the adhesion of the deposited film to the substrate [359]. Plasmas can also generate electronic sites on ceramics that can affect nucleation and adhesion [360].

The first studies on plasma treatment of polymers were for adhesive bonding. The first studies used corona discharges in air by Rossman [361] in 1956. In the late 1960s inert gas plasmas were used in the process called "crosslinking by activated species of inert gas" ("CASING") [362], though the inert gas was probably contaminated by reactive species. Plasma treatment can also be used to crosslink low-molecular species on the surface, thus avoiding the "weak boundary layer" adhesion problem. UV radiation from the plasma probably plays an important role in this "curing" process.

A substrate surface can be hardened by the formation of dispersions of carbide, nitride, or boride phases in the near-surface region. This can be done external to the deposition system by diffusion of reactive species into the appropriate alloy material at high temperature. Dispersion hardening can be done in vacuum by ion implantation of one (example: N) or several (example: Ti + C) species [363] or in a plasma by ion bombardment and heating by the reactive species [344]. The term "ionitriding" is used for the plasma nitriding process [364] and "plasma carburizing" for forming a carbide phase [365]. In "plasma source ion implantation," pulsed plasmas are used to diffuse reactive species into the surface [339].

Vapor deposited fluid films are used to "flow coat" webs in vacuum to cover surface defects and smooth the surface. These polymer films are then cured with UV or electron beam irradiation [214].

The chemistry of a surface can be changed by depositing a "glue layer" such as in the glass-Ti-Au system where the titanium is oxygen-active and reacts with the glass, and gold is soluble in the titanium [348]. In reactive deposition the intermediate layer can be deposited by limiting the availability of the reactive species. For example, the system may be: substrate-Ti-TiN$_{x-1}$-TiN, formed by limiting the availability of the nitrogen. This is called a "graded" interface [348].

Recontamination of surfaces in the deposition chamber can be a major problem. In addition to the obvious problems (of backstreaming, flaking, and wear particles), particles can be formed by vapor phase nucleation of the vaporized material [196, 366, 370]. If there are vapors in the system rapid evacuation can cause cooling and vapor phase condensation of droplets that can "rain" on the substrate surfaces. This effect was reported in 1990 [367], but the effect of cloud" formation by rapid expansion and cooling had been reported by C.T.R. Wilson in 1928 and was the basis of the 1948 Nobel Prize by P.M.S. Blackett for work using the "Wilson cloud track chamber."

The preparation of "vacuum surfaces" in the deposition chamber is important in generating a clean vacuum environment rapidly [368]. Both oxygen [358] and hydrogen [357] plasmas have been used to clean vacuum surfaces. Reactive plasma cleaning is used for *in situ* cleaning of vacuum surfaces in vacuum systems and to "strip" deposits from vacuum surfaces.

Stripping and reclamation of substrates, fixtures, and liners can be important to economical vacuum coating [368a]. For example, astronomical telescope mirrors are routinely stripped and recoated [150].

Summary

Today there are thousands of applications of vacuum coating technologies. It is hard to realize that the industry is only about 70 years old. The vapor sources for vacuum coating are 100 to 150 years old but commercial uses did not start until the mid-1930s with the development of thermal evaporation in vacuum. The applications of vacuum coatings progressed from the simple single-layer coatings used for electrical, optical, and reflecting applications in the 1930s and 1940s to coatings for corrosion protection in the 1950s. In the 1950s coatings on flexible materials for packaging began to be used and "vacuum metallization" for decorative purposes became a big business. The advent of semiconductor technology and the need for electrically conductive metallization and passivation layers was a major impetus to vacuum coating. The "energy crisis" in the 1970s showed the need for energy-conservation coatings on large areas of glass and polymer webs. In the 1980s vacuum coatings for display applications, particularly transparent conductive oxides, became important, and in the 1990s hard coatings for tools and decorative applications became important new applications. In the future it is expected that vacuum coatings will continue to play a vital role in developing both existing and new products. An example are the optically variable interference/diffraction films that are fractured and used as pigments in ink to counter counterfeiting [369].

As with many inventions and inventors [370] (E-28) it is often difficult to establish the originator of any particular idea or invention. The search is complicated by the information being in both the patent and the technical literature. Patents seldom cite relevant technical papers and technical papers seldom cite relevant patent literature. Often there is controversy over aspects of the information provided.

The sputtering phenomenon was widely studied in the 1920s and 1930s (Guentherschulze, et al) as well as in the 1950s and 1960s (Wehner). Sputter deposition was the first vacuum coating technology to be available, but with the exception of Edison's deposition on thermally sensitive wax substrates and some work on specialized mirrors, the technique was not widely used until the advent of semiconductor device fabrication. The use of encapsulated aluminum metallization on silicon devices revealed the problem of "stress voiding" that could be solved by the use of an Al-Cu-Si alloy metallization [371]. Sputter deposition was the answer [372].

Applications of sputter deposition increased rapidly after the invention of the various high-rate magnetron sputtering sources in the early 1970s. Reactive sputter deposition of electrically conductive compound materials (such as the nitrides and the carbides) for decorative and hard-coating applications began in the early 1980s. Target "poisoning" and surface flashover (arcing) were problems when reactive sputtering "good" dielectric materials, such as most oxides. The invention of the dual magnetron and pulsed-power sputtering sources in the early 1990s helped alleviate those problems.

Thermal evaporation was an obvious vapor source long before it was studied. Its development was inhibited by the high radiant-heat loads and the lack of vacuum materials and techniques that could withstand the heat, particularly in a demountable system. As John Strong comments in his book, "Although the evaporation method was known by 1912, it remained obscure, for some reason, long after it should have become a practical 'tool' in the laboratory" [373]. Thermal evaporation began to be developed in earnest after the work of John Strong on the aluminization of astronomical mirrors in the mid-1930s. Aluminum was one desirable material that could not be electroplated without great danger and difficulty. The technology advanced further with the development of e-beam evaporation that allowed refractory materials to be deposited. Evaporation into plasmas extended the technique into the realm of practical reactive evaporation.

Arc vapor deposition had sporadic development in the 1970s but really became important in the 1980s with the introduction of the reactive arc deposition work of the Russians. Arc vaporization is less sensitive to source "poisoning" than is sputtering and has the advantage that much of the vaporized material is ionized. The application of arc deposition became more useful as extended arc sources, such as rod and elongated "racetrack" sources, became available.

Thermally driven chemical vapor deposition (CVD) had limited usefulness because of the high temperatures required. High temperature in general requires that the coefficient of thermal expansion of the substrate and the coating must be matched, or else excessive strains are introduced into the coating. The advent of plasma-enhanced CVD lowered the deposition temperature required and PECVD began to be used, particularly in the semiconductor processing industry. At the pressures used for PECVD it was not possible to accelerate ions to high energies. This was overcome by the use of LP-CVD, which was done at "sputtering pressures" (i.e., 10 mTorr or so).

Ion plating utilizes continuous or periodic bombardment by high-energy atomic-sized particles, generally ions, to create a "clean" interface and to modify the properties of the deposited material. The bombardment densifies the deposited material and enhances the chemical reactions necessary for reactive deposition. The concept of bombardment during deposition was patented in the late 1930s but was "lost" until the early 1960s. At that time the process was patented using chemical vapor precursors (LP-CVD) and thermal evaporation sources. Sputter deposition at that time was slow and the use of "bias sputtering" (bias sputter deposition) was lim-

ited. The advent of magnetron sputtering made the use of concurrent bombardment during sputter deposition more attractive.

Of course it is impossible to list all of the relevant references but the author hopes that he has correctly identified the "first" reference to important vacuum coating processes and process parameters and has included enough later references to help the reader "fill in the blanks" where so desired. It is of interest to note that one of the first commercial vacuum coating companies in the USA was established in 1936 (E-29).

The author has not attempted to identify all or even most of the applications of vacuum coating and the work that has been done on characterizing their growth and properties. He would, however, like to point out that "there are no 'handbook values' for the properties of vacuum coatings—their properties depend strongly on the substrate properties (particularly surface roughness), the deposition process, and the processing parameters used to create them."

Endnotes

(E-1) To be a "pioneering" work as far as an application is concerned, the author means that the work was pursued with "vigor" by the initial investigator or other contemporaries, and that the initial work was recognized by future investigators. The key seems to be that there was a need or there was a marketing operation capable of creating a perceived need, otherwise the work was "lost."

(E-2) Abbe Jean-Antonne Nollet was a famous French scientist studying electrostatic ("frictional") electricity. In 1751 Nollet sent an electrical shock, from a bank of Leyden jars, through 180 soldiers holding hands and made them jump before of the King of France. Later the King had Nollet make 700 Carthusian monks jump with an electrical shock.

(E-3) Ben Franklin was the prime mover in bringing the French into the Revolutionary War on the side of the Americans. This act was a major factor in the withdrawal of the British from what became the United States of America. One of the major reasons that Mr. Franklin was accepted by the French Court in1776 was that he was an established scientist who was a Fellow of the Royal Society (London), and the French Court appreciated scientists.

(E-4) At the turn of the century Thomas A. Edison was an advocate of "safe" DC power while George Westinghouse was promoting AC power. Edison designed and built an electric chair using AC and made a movie of a simulated electrocution of a man in 1902 to show how dangerous AC was. He routinely electrocuted dogs and cats in public to demonstrate the dangers of AC and in 1903 he electrocuted a rogue zoo elephant named Topsy. See Richard Moran, *Executioner's Current: Thomas Edison, George Westinghouse, and the Invention of the Electric Chair*, Knopf (2002)

(E-5) One of the first commercial applications of hydroelectric AC was in the mining town of Telluride, Colorado. In 1891, a water-powered Westinghouse AC generator produced power that was transmitted 2.5 miles to power a 100 horsepower AC induction (Tesla) motor at the Gold King mine. This hydroelectric power replaced the steam-generated DC electric power that had become too expensive. The cost of electricity at the mine dropped from $2,500/month to $500/month. Several years later the town of Telluride was electrified. This installation preceded the Niagara Falls hydroelectric AC generating plant that came on-line in 1895.

(E-6) The introduction and use of "pulsed power" brought forth a number of terms such as "asymmetrical AC," "unipolar pulsed DC," "bipolar pulsed DC," "counterpulse," and even "bipolar DC." These terms were often not defined by the authors and sometimes flew in the face of accepted usage of electrical terminology.

(E-7) Professor Brown tells the story of Hittorf who labored week-after-week, gradually extending the length of a thin glass discharge tube trying to discover the length of the positive column. Eventually the tube ran back and forth across Hitterof's laboratory. At that stage, a frightened cat, pursued by a pack of dogs, came flying through the window... "Until an unfortunate accident terminated my experiment," Hittorf wrote, "the positive column appeared to extend without limit." [ref. 22 p.77]

(E-8) Sir William Crookes was an eminent scientist but raised a lot of controversy when he "experimentally verified" that several psychic mediums, including Daniel Dunglas Homes and Florence Cook, were authentic and that their abilities "transcended the laws of nature." Scientific journals refused to publish many of his "findings" and much of his work on that subject was privately published. Ref: "Sir William Crookes—FRS, FRGS, *Researches in the Phenomena of Spiritualism*, Pantheon Press (1971) (reprints of many of Crookes' articles).

(E-9) In this time period the term "arts" was used as we use the term "technology" today. The term is still used in the patent-related literature, e.g., "prior art."

(E-10) "I find in practice that the employment of an electric arc for vaporizing the metal, as suggested in my patent, is open to the objection of being slow, and unless the process is carried out with great care the deposit is not entirely uniform, while there is danger of injuring the very delicate phonograph record surface, particularly from the heat of the arc. I find that the rapidity of the process is increased and the character of the deposit improved if the vaporization of the metal is effected by maintaining between two electrodes of the metal a silent discharge of electricity of high tension, such as may be produced from an induction-coil of large capacity or from any induction-machine of approved type, such as the Helmholtz induction-machine." (From ref. [42].)

(E-11) The question arises as to why much of the early

work done under government contract or by government employees was not patented. Before 1980 the U.S. government employees and contractors were only interested in putting R&D work in the Public Domain because they were not allowed to patent and license the work for profit. In 1980 Congress passed legislation (the Bayh-Dole Act and the Stevenson-Wydler Technology Innovation Act) that allowed work developed with government support to be patented and protected. Patents then became much more desirable at government-owned government-operated (GOGO) and government-owned contractor-operated (GOCO) establishments such as the Rocky Flats Plant. However, the initial work on magnetron sputtering from a hemispherical target was for a classified project.

(E-12)   In 1970 John Chapin, John Mullay, and Ted van Vorous—all of whom had worked at the Dow Chemical Co., Rocky Flats Plant—formed Vacuum Technology Associates (VacTec). VacTec had a contract with Airco to develop a long, linear, high-rate sputtering source. Airco provided money and equipment for the development work. Chapin initially used a trough-shaped source similar in cross-section to the hemispherical configuration that Mullaly had used in 1969. He quickly obtained the desired sputtering rate for copper. He also conceived the flat planar design. Luckily he noted the idea in his engineering notebook and that entry was witnessed by Bob Cormia of Airco. It would seem that Corbani's patent "anticipated" Chapin's patent. However Chapin was able to "swear behind" Corbani's disclosure based on the notebook entry and therefore Chapin's patent took precedence and Chapin is credited with inventing the planar magnetron sputtering source. Initially there was disagreement between Airco and VacTec about who owned the patent for the planar magnetron. The disagreement was played out in court with the decision in favor of Airco for long sources. Afterward Airco made and marketed long planar magnetron sources and VacTec made shorter versions.

(E-13)   Shorter Oxford Dictionary, 3rd edition (1955)—Sputter v (1598) or sputter sb (1673) also splutter sb (1823)—an English term meaning "To spit out a spray of particles in noisy bursts." Imitative origin from the Dutch word sputteren.

(E-14)   L. Holland has the wrong page [p. 546] for ref. 115 in his book (107, reference 2) and this mistake has been perpetuated in several subsequent works.

(E-15)   John Strong gives 1928 as the date of Ritschl's work in the text of his book (ref. 131, p. 171), but gives ref. 144 as the reference to Ritschl's work. Strong could be called the "father of filament evaporation" for his long efforts in this area, and he was a contemporary of Ritschl so the author thinks that he must have had a good reason for using 1928 as the year of the work.

(E-16)   Before the use of aluminizing to coat astronomical front-surface mirrors, chemically deposited silver was used [40]. The silver coating had to be carefully polished but polishing left minute scratches that limited the resolution of the mirror. For example, with a silver coating on the 100 inch Crossley reflector telescope at the Lick Observatory, the companion star to Sirius was very difficult to resolve. Using a thin (1000 Å) aluminum coating, the companion was easily seen [141]. Silver also tarnished with age, reducing the reflectivity. Aluminum, on the other hand, forms a protective oxide coating and the reflectivity remains constant. In some cases an SiO coating is applied for additional protection.

(E-16a) Possibly the earliest known example of an aluminized astronomical mirror by John Donovan Strong is on display at the Chabot Space and Science Center, Oakland, CA. It is in their exhibit "Astronomy in California 1850-1950: Telescope Makers, Telescopes and Artifacts." The mirror is a glass parabola of 6 5/32 inch aperture with a front surface aluminum metallization.

(E-17)   "One of the men working for Wright was an amateur astronomer who had trouble with silver wearing off his telescopic mirror. In their lab was a machine that could vaporize aluminum so they tried coating one of his mirrors in 'that vaporizing machine.' Wright was so inspired by the results that he immediately began work on a reflector lamp. A glass custard cup was purchased at the dime store, coated with aluminum and a filament rigged up. For the lens a curved section was cut out of a giant incandescent lamp. It didn't take long to realize his idea was indeed a good one." (Quoted from [129]).

(E-18)   After WWII the Japanese camera makers (Canon and Nikon) infringed on many German camera patents. When the Germans complained, the Allied Control Commissions for both Germany and Japan took no action. This allowed the Japanese to rapidly build up their camera industry to the dismay of the Germans. (Information from "Post War Camera & Lens Design Thievery" by Marc James Small <teachnet.edb.utexas.edu/~leica/thievery>). As far as the author can tell the Japanese did not use coated optics during WWII.

(E-19) There is some question about the amount of gas phase nucleation that occurred in Takagi's nozzle-expansion technique. It would seem that many of the effects reported by Takagi were due to accelerated ions, not accelerated atomic clusters.

(E-20) When an American jeweler, J. Filner, who was visiting the USSR, noted the gold color of arc-vapor-deposited TiN, he brought the process to the United States in about 1980. (From [38]).

(E-21) The "Calutron" was named for the University of California where the source was developed. The source used a confined plasma to dissociate UCl_4 to provide metal ions of U^{235} and U^{238}, which is the dominant uranium isotope. The ions of the isotopes were then magnetically separated to give "enriched" and "weapons grade" (93.3% U^{235}) uranium. The design of this source was considered classified for many years after WWII.

(E-22) Mattox conceived of the ion plating process while discussing the adhesion of metal films to metal with proponents of the theory that diffusion was necessary for good adhesion. Mattox maintained that a clean surface (interface) was a sufficient criterion. Mattox knew that sputter cleaning was being used to produce atomically clean surfaces in ultra-high vacuum for low-energy electron-diffraction (LEED) studies. Using what became the ion plating process, Mattox demonstrated good adhesion between silver films and iron substrates—iron and silver have no solubility even in the molten state. Later silver (a low-shear-strength metal) on steel bearings for solid-film tribological use became an important application of ion plating.

(E-22a) Mattox first reported on the ion plating process at the summer Gordon Research Conference on Adhesion in 1963. His paper was scheduled for Thursday evening after the all-you-can-eat lobster dinner. It was not until Wednesday that he received clearance from the Sandia Patent Office to present the paper. After the lobster dinner he presented the paper and the paper and subsequent discussion ran on for over three hours! Mattox often said it was one of the most difficult papers that he ever presented.

(E-23) Ivadizer™ is the trademark of the McDonnell Douglas Corp., which marketed the equipment for many years as well as coating production parts themselves.

(E-24) The term "diamond-like carbon" (DLC) was first used by S. Aisenberg in 1972 [271]. Some authors use this term to describe hard carbon films with a low hydrogen content, and the term "diamond-like hydrocarbon" (H-DLC) films for hard carbon films containing appreciable hydrogen [E-24-1]. The term "i-C" has also been used for ionized and accelerated carbon ions [E-24-2]. The term "a-C" is used for amorphous carbon. Metal-containing DLC coatings were developed in the mid 1980s for tribological coatings [E-24-3] and are designated Me-DLC. Metal-containing MoS_2 coatings had previously been used in tribological applications.

[E-24-1] "Carbon Thin Films," J.C. Angus, P. Koidl, and S. Domitz, p. 89 in *Plasma Deposited Thin Films*, edited by J. Mort and F. Hansen, CRC Press (1986)

[E-24-2] C. Weismantel, C. Schürer, R. Frohlich, P. Grau, and H. Lehmann, *Thin Solid Films*, 61, L5 (1979)

[E-24-3] H. Dimigen, H. Hübsch, and R. Memming, "Tribological and Electrical Properties of Metal-Containing Hydrogenated Carbon Films," *Appl. Phys. Lett.*, 50, 1056 (1987)

(E-25) One of the first experiments on using UV/O_3 cleaning involved using a "black light" purchased at a gem and mineral store. A clean glass slide (clean as determined by a "water drop wetting angle" test) was placed on a stainless steel bench top in the lab. The black light (UV source) was placed over it and the wetting angle monitored for several weeks. No change was found. The control slide, with no black light, showed wetting angle increases within the first hour. Recontamination was by adsorption of vapors from the atmosphere.

(E-26) "Clean bench" or "clean room" is somewhat of a misnomer in that the air is generally only cleaned of particles and **not** vapors. In rare cases activated carbon filters are also used to remove vapors.

(E-27) When preparing to aluminize the Palomar mirror, John Strong notified the mirror polishers that he would be using a new cleaning technique using "a special fatty acid compound with precipitated chalk." When he arrived the "special fatty acid compound" was Wild Root Cream Oil hair tonic (ad jingle—"You better get Wild Root Cream Oil, Charlie; It keeps your hair in trim; Because it's non-alcoholic, Charlie; It's made with soothing lanolin"). He stated, "In order to get glass clean you first have to get it properly dirty." The oil residue was "burned-off" using an oxygen plasma in the vacuum deposition chamber. (From [149].)

(E-28) Rev. Hannibal Goodwin was a pastor who developed the flexible photographic film in 1887 to replace glass-plate negatives and make it easier to photograph travelogues to holy sites. Eastman

Kodak worked to prevent his patenting the idea until 1898 when Rev. Goodwin was granted his patent. After that Kodak infringed on the patent and the litigation was not settled until after WWI in favor of Ansco ($5M) who had bought the patent rights after Goodwin was killed in a street accident in 1900.

(E-29) In 1936 the company Evaporated Metal Films was formed in Ithaca, New York, by Robley Williams, John Ruedy, and Joel Ufford, all of Cornell University. This company is still in business as a privately held company and is probably the oldest "contract coating" firm performing vacuum coating in the USA.

References

1. *Vapor Deposition*, edited by Carrol F. Powell, Joseph H. Oxley, and John M. Blocher, Jr., John Wiley (1966)

2. *Vacuum Deposition of Thin Films*, L. Holland, Chapman Hall (1956)

3. H.-D. Steffens, H.-M. Hohle, and E. Erturk, "Low Pressure Plasma Spraying of Reactive Materials," *Thin Solid Films*, 73, 19 (1980)

4. *Handbook of Thick Film Technology*, P.J. Holmes and R.G. Loasby, Electrochemical Publication Ltd. (1976)

5. *History of Vacuum Science and Technology*, edited by Theodore E. Madey and William C. Brown (American Vacuum Society), American Institute of Physics (1984)

6. *Vacuum Science and Technology: Pioneers of the 20th Century*, edited by Paul Redhead, AIP Press (1994)

7. *Adventures in Vacuum*, M.J. Sparnaay, North-Holland (1992)

8. R.K. Waits, "Vacuum Through the Years," *Vacuum & Thin Films*, 2(1) 32 (1999)

9. http://www.iuvsta.org/vsd/biblio.htm—extensive bibliography of vacuum science and technology texts; also The Belljar, http://www.tiac.net/users/shansen/belljar/

10. Georgius Agricola, *De Re Metallica* (Secrets of Mining & Refining), (1556), (translated by Herbert Hoover and Lou Henry Hoover), Dover Publications (1950)

11. C.R. Meissner, "Liquid Nitrogen Cold Traps," *Rev. Sci. Instrum.*, 26, 305 (1955); also *Vacuum Technology and Space Simulation*, Donald J. Santeler, Donald W. Jones, David H. Holkeboer, and Frank Pagano, NASA SP-105, NASA (1966); also *Capture Pumping Technology*, Kimo M. Welch, Elsevier Science (2001)

12. R.K. Waits, "Evolution of Integrated-Circuit Vacuum Processes: 1959–1975," *J. Vac. Sci. Technol.*, A18(4) 1736 (2000)

12a. J.G. Simmons and L.I. Maissel, "Multiple Cathode Sputtering System," *Rev. Sci. Instrum.*, 32, 542 [1961]; also L.I. Maissel and J.H. Vaughn, "Techniques for Sputtering Single and Multilayer Films of Uniform Resistivity," *Vacuum* 13, 421 (1963)

12b. Sidney S. Charschan and Harald Westgaard, "Apparatus for Processing Materials in a Controlled Atmosphere," U.S. Patent 3,294,670 (Dec. 27, 1966)

12c. W.L. Shockley, E.L. Geissinger, and L.A. Svach, "Automatic Sputtering of Tantalum Films for Resistor and Capacitor Fabrication," *IEEE Transactions on Components and Parts,* CP11, 34 (1964); also *Thin Film Technology,* Robert W. Berry, Peter M. Hall, and Murray T. Harris, p. 250, Van Nostrand Reinhold (1968)

12d. Joseph S. Mathias, Alfred A. Adomines, Richard H. Storck and John McNamara, "Evaporation System," U.S. Patent 3,404,661 (Oct. 8, 1968); also William C. Lester and Ernest S. Ward, "Continuous Vacuum Process Apparatus," U.S. Patent 3,652,444 (Feb. 28, (1972)

13. *Silicon Processing for the VLSI Era: Vol. 1: Process Technology* (second edition), S. Wolf and R.N. Tauber, Section 6.2.5.3 (2000)

13a. H.R. Smith and C.d'A. Hunt, "Methods of Continuous High Vacuum Strip Processing," p. 227 in *Transactions of The Vacuum Metallurgy Conference*, American Vacuum Society (1964)

14. D.M. Mattox, "Steady State and 'Transit' Conduction," *Vac. Technol. Coat.*, 2(6) 20 (2001); also C. Hayashi, "The Role of Adsorption in Production and Measurement of High Vacuum," *Vacuum Technology Transactions—Proceeding of the Fourth National Symposium*, American Vacuum Society, p. 13, Pergamon Press (1957)

15. *Vacuum Arcs: Theory and Practice*, edited by J.W. Lafferty, John Wiley (1980)

16. *Alessandro Volta and the Electric Battery*, Bern Dibner, Franklin Watts, Inc. (1964)

17. *A History of Electricity and Magnetism*, Herbert W. Meyer, Brundy Library (MIT) (1971)

18. *Benjamin Franklin, Electrician, Bern Dibner, Brundy Library (MIT) (1971)*

19. *The Founders of Electrochemistry*, Samuel Ruben, Dorrance & Co. (1975)

20. *The Birth of the Vacuum Tube: The Edison Effect*, Fernand E. d'Humy, Newcomen Society (1949); also *The Edison Effect*, Harold G. Bowen, The Thomas Alva Edison Foundation (1951); also *The Edison Effect*, Ron Ploof, Cypress Publications (1995)

21. Gunter Mark, "Low Frequency Pulsed Bipolar Power Supply for a Plasma Chamber," U.S. Patent #5,303,139 (April 12, 1994)

22. J.E. Brittain, "The Magnetron and the Beginnings of the Microwave Age," *Physics Today*, p. 60 (July 1985)

23. *The Electrochemistry of Gases and Other Dielectrics*, G. Glocker and S.C. Lind, John Wiley & Sons (1939)

24. *Plasma Chemistry in Electrical Discharges*, F.K. McTaggart, Elsevier (1967); also S. Veprek, "Plasma-Induced and Plasma-Assisted Chemical Vapour Deposition," *Thin Solid Films* 130, 135 (1985); also S.C. Brown in *Gaseous Electronics*, edited by J.W. McGowan and P.K. John, North-Holland Publishing (1974)

25. *Glow Discharge Processes*, Brian Chapman, Wiley-Interscience (1980); also *Cold Plasmas in Materials Fabrication: From Fundamentals to Applications*, Alfred Grill, IEEE Publications (1994)

26. "The Historical Development of Controlled Ion-Assisted and Plasma-Assisted PVD Processes," D.M. Mattox, p. 109 in *Proceedings of the 40th Annual Technical Conference*, Society of Vacuum Coaters (1997)

27. *Fundamentals of Gaseous Ionization and Plasma Electronics*, E. Nasser, Wiley Interscience (1972)

28. D.E. Hull, "Induction Plasma Tube," U.S. Patent #4,431,901 (Feb. 14, 1984)

29. P.D. Reader, "Ion Beam Sources, Past, Present and Future," p. 3 in *Proceedings of the 42nd Annual Technical Conference*, Society of Vacuum Coaters (1999); also H.R. Kaufman, J.J. Cuomo, and J.M.E. Harper, "Technology and Application of Broad-Beam Ion Sources Used in Sputtering: Part 1—Ion Source Technology," *J. Vac. Sci. Technol.*, 21(3) 725 (1982); also J.E. Keem, "The History, Theory, and Application of Ion Sources," *Vac. Technol Coat.*, 3(9) 46 (2002)

29a. J.D. Gow and L. Ruby, "Simple, Pulsed Neutron Source Based on Cross-Field Trapping," *Rev. Sci. Instrum.*, 30(5) 315 (1959)

30. N.V. Filipova, T.I. Filipova, and V.P. Vingredov, *Nuclear Fusion Suppl.*, 2, 577 (1962); also J.W. Mather, *Physics of Fluids Suppl.*, 7, 5 (1964)

31. J.M. Lafferty, "History of the International Union for Vacuum Science, Technique, and Applications: Review Article," *J. Vac. Sci. Technol.*, A5(4) 405 (1987)

32. United States Court of Appeals for the Federal Circuit, Opinion 00-1241, Litton Systems Inc. v. Honeywell Inc. (Decided: February 5, 2001); also Litton Sys., Inc. v. Honeywell, Inc. 87 F.3d 1559, 39 USPQ2d 1321 (Fed. Cir. 1996); also United States Court of Appeals for the Federal Circuit 95-1242-1311 (Decided: July 3, 1996)

33. *Handbook of Plasma Processing Technology*, edited by Stephen M. Rossnagel, Jerome J. Cuomo, and William D. Westwood, William Andrew Publishing/Noyes Publications (1990)

34. *Thin Film Processes*, edited by John L. Vossen and Werner Kern, Academic Press (1979)

35. W.R. Grove, "On the Electrochemical Polarity of Gases," *Phil. Trans. Royal. Soc. (London),* B142, 87 (1852)

36. J. Plücker, "Observations on the Electrical Discharge Through Rarefied Gases," *The London, Edinburgh and Dublin Philosophical Magazine*, 16, 409 (1858)

37. A.W. Wright, "On the Production of Transparent Metallic Films by Electrical Discharge in Exhausted Tubes," *Am. J. Sci. Arts*, Vol. 13, pp. 49-55 (1877); also "On the New Process for the Electrical Deposition of Metals and for Constructing Metal-Covered Glass Specula," *Am. J. Sci. Arts*, Vol. 14, pp. 169-178 (1878)

38. "Vacuum Arc Deposition: Early History and Recent Developments," Raymond L. Boxman, Invited Dyke Award and Lecture, pp. 1-8 in *Proceedings of XIXth Symposium on Discharges and Electrical Insulation in Vacuum,* (IEEE), Xi'an, China, September 18–20, 2000

39. D.M. Mattox, "Sputter Deposition: Early Efforts," *Plat. Surf. Finish.*, 88(9) 60 (2001); also *Vac. Technol. Coat.*, 2(9) 34 (July 2001)

40. T.A. Edison, "Art of Plating One Material on Another," U.S. Patent 526,147 (filed 1884) (Sept. 18, 1894)

41. T.A. Edison, "Process of Duplicating Phonograms," U.S. Patent #484,582 (Oct. 18, 1892); also T.A. Edison, U.S. Patent # 526,147 (Sept. 18, 1894)

42. T.A. Edison, "Process of Coating Phonograph-Records," U.S. Patent #713,863 (Nov. 18, 1902)

42a. G. Herdt, A. McTeer and S. Meikle, "PVD Copper Barrier/Seed Processes: Some Considerations for the 0.15 um Generation and Beyond," *Semiconductor Fabtech.* 11, 259 (1999).

43. T.A. Edison, U.S. Patent #1,163,329 (Dec. 1915)

44. W. Crookes, "On Electrical Evaporation," *Scientific American Supplement*, Vol. 32 (811) pp. 12958–12960, July 18 (1891); also *Proc. Roy. Soc. (London)*, 50, 88 (1891)

45. "Note on the Production of Mirrors by Cathodic Bombardment," F. Simeon, p. 26 in *The Making of Reflecting Surfaces*, a discussion held by the Physical Society of London and the Optical Society (Nov. 26, 1920), Fleetway Press, Ltd. The text may be found in the History section of the Society of Vacuum Coaters web site, www.svc.org

46. V. Kohlschütter, *Jahrb. d. Radioactivität*, 9, 355 (1912)

47. O. Almen and G. Bruce, *Nucl. Instrum. Meth.*, 11, 279 (1961)

48. J.S. Colligon, C.M. Hicks, and A.P. Neokleous, "Variation of the Sputtering Yield of Gold with Ion Dose," *Rad. Effects,* 18, 119 (1973); also J.S. Colligon and M.H. Patel, "Dependence of Sputtering Coefficient on Ion Dose," *Rad. Effects*, 32, 193 (1977)

49. E.V. Kornelsen, "The Interaction of Injected Helium with Lattice Defects in a Tungsten Crystal," *Rad. Effects*, 13, 227 (1972); also D. Chleck, R. Maehl, O. Cucchiara, and E. Carnevale, *Int. J. Appl. Radiation Isotopes*, 14, 581 (1963)

50. J.P. van der Slice, "Ion Energies at the Cathode of a Glow Discharge," *Phys. Rev.,* 131, 219 (1963); also J. Machet, P. Saulnier, J. Ezquerra, and J. Gulle, "Ion Energy Distribution in Ion Plating," *Vacuum*, 33, 279 (1983); also G.J. Kominiak and J.E. Uhl, "Substrate Surface Contamination from Dark-Space Shielding During Sputter Cleaning," *J. Vac. Sci. Technol.*, 13(6) 1193 (1976)

51. I.C. Gardner and F.A. Case, *The Making of Mirrors by the Deposition of Metals on Glass,*" Bureau of Standards, Circular #389 (January 1931) (available from Lindsay Publications as "How to Make Mirrors" which does not give the names of the original authors)

52. "Vacuum Web Coating—An Old Technology With a High Potential for the Future," E.O. Dietrich, R. Ludwig, and E.K. Hartwig, p. 354 in *Proceedings of the 40th Annual Technical Conference,* Society of Vacuum Coaters (1997)

53. C.J. Overbeck, *J. Opt. Soc. Am.*, 23, 109 (1933); also H. Köenig and G. Helwig, *Optik,* 7, 294 (1950)

54. G.A. Veszi, *J. Brit. Instn. Radio Eng.*, 13, 183 (1953)

55. D. Gerstenberg and C.J. Calbrick, "Effects of Nitrogen, Methane and Oxygen on the Structure and Electrical Properties of Thin Tantalum Films," *J. Appl. Phys.*, 35, 402 (1964); also *Thin Film Technology,* Robert W. Berry, Peter M. Hall, and Murray T. Harris, Van Nostrand Reinhold (1968)

56. R.W. Berry and D.J. Sloan, "Tantalum Printed Capacitors," *Proc. of IRE*, 47, 1070 (1959)

57. F.M. Penning, "Coating by Cathode Disintegration," (filed December 1935 in Germany), U.S. Patent #2,146,025 (Feb. 7, 1939); also *Physica (Utrecht),* 3, 873 (1936)

58. F.M. Penning and J.H.A. Mobius, *Proc., K. Ned. Akad. Weten.*, 43, 41 (1940): also A.S. Penfold, *Thin Solid Films*, 171, 99 (1989)

59. J.A. Thornton and A.S. Penfold "Cylindrical Magnetron Sputtering," Sec. II-2 in *Thin Film Processes*, edited by J.L. Vossen and Werner Kern, Academic Press (1978); also J.A. Thornton, *J. Vac. Sci. Technol.*,15, 171 (1978); also A.S. Penfold and J.A. Thornton, U.S. Patents #3,884,793 (1975), #3,995,187 (1976), #4,030,996 (1977), #4,031,424 (1977), #4,041,053 (1977), #4,111,782 (1978), #4,116,793 (1978), #4,116,794 (1978), #4,132,612 (1979), #4,132,613 (1979), and A.S. Penfold, #3,919,678 (1975)

60. D.M. Mattox, R.E. Cuthrell, C.R. Peeples, and P.L. Dreike, "Design and Performance of a Moveable-Post Cathode Magnetron Sputtering System for Making PBFA II Accelerator Ion Sources," *Surf. Coat. Technol.*, 33, 425 (1987)

61. "High Rate Sputtering with Torous Plasmatron," U. Heisig, K. Goedicke, and S. Schiller, p. 129 in *Proceedings of the 7th International Symposium on Electron and Ion Beam Science and Technology*, Electrochemical Soc. (1976); also J.A. Thornton and V.L. Hedgcoth, "Tubular Hollow Cathode Sputtering onto Substrates of Complex Shapes," *J. Vac. Sci. Technol.*, 12(1) 93 (1975)

62. "Principles and Application of Hollow Cathode Magnetron Sputtering Sources," D.A. Glocker, p. 298 in *Proceedings of the 38th Annual Technical Conference,* Society of Vacuum Coaters (1995); also D.E. Diegfried, D. Cook, and D.A. Glocker, "Reactive Cylindrical Magnetron Deposition of Titanium Nitride and Zirconium Nitride Films," p. 97 in *Proceedings of the 39th Annual Technical Conference,* Society of Vacuum Coaters (1996)

63. G.S. Anderson, W.N. Mayer, and G.K. Wehner, *J. Appl. Phys.,* 33, 2991 (1962); also G.S. Anderson and R.M. Moseson, U.S. Patent #3,233,137 (Nov. 1966)

64. "Sputtering by Ion Bombardment," G.K. Wehner, p. 239 in *Advances in Electronics and Electron Physics*, Vol. VII, Academic Press (1955)

65. J.K. Robertson and C.W. Clapp, *Nature*, 132, 479 (1933)

66. P.D. Davidse and L.I. Maissel, *J. Appl. Phys.*, 37, 574 (1966); also *Vacuum*, 17, 139 (1967); also "Apparatus for Cathode Sputtering Including a Shielded RF Electrode," (filed Jan. 28, 1965), U.S. Patent #3,369,991 (Feb. 20, 1968)

67. J. Hohenstein, "Cermet Resistors by Concurrent RF and dc Sputtering," *J. Vac. Sci. Technol.*, 5(2) 65 (1968)

68. H.S. Butler and G.S. Kino, *Phys. Fluids*, 6, 1346 (1963)

69. G.K. Wehner, "Growth of Solid Layers on Substrates Which are Kept Under Ion Bombardment Before and During Deposition," (filed April 27, 1959), U.S. Patent #3,021,271 (Feb. 1962)

70. L.I. Maissel and P.M. Schaible, "Thin Films Deposited by Bias Sputtering," *J. Appl. Phys.*, 36, 237 (1965); also O. Christensen, "Characteristics and Applications of Bias Sputtering," *Solid State Technol.*, 13, 39 (Dec. 1970)

71. H.C. Cooke, C.W. Covington, and J.F. Libsch, "The Preparation and Properties of Sputtered Aluminum Thin Films," *Trans. Metallurgical Soc.*, AIME, 236, 314 (1966)

72. F.M. d'Heurle, "Resistivity and Structure of Sputtered Molybdenum Films," *Trans. Metallurgical Soc.*, AIME, 236, 321 (1966)

73. R.D. Ivanov, G.V. Spivak, and G.K. Kislova, *Izv. Akad. Nauk SSSR Ser. Fiz.*, 25, 1524 (1961); also J. Edgecumbe, L.G. Rosner, and D.E. Anderson, *J. Appl. Phys.*, 35, 2198 (1964); also "Application of Sputtering to the Deposition of Films," Leon Maissel, Ch. 4, p. 4–8 in *Handbook of Thin Film Technology*, edited by Leon I. Maissel and Reinhardt Glang, McGraw-Hill Publishers (1970)

74. T.C. Tisone, "Low Voltage Triode Sputtering with a Confined Plasma," *J. Vac. Sci. Technol.*, 12(5) 1058 (1975)

75. W.W.Y. Lee and D. Oblas, *J. Appl. Phys.*, 46, 1728 (1975); also H.S.W. Massey and E.H.S. Burhop, *Electronic and Ionic Impact Phenomena*, Oxford University Press (1952); also H.G. Hagstrum, p. 1 in *Inelastic Ion Surface Collisions*, edited by N.H. Tolk, J.C. Tully, W. Heiland, and C.W. White, Academic Press (1977); also W. Molthan, *Z. Physik*, 98, 227 (1936)

76. R.E. Jones, C.L. Standley, and L.I. Maissel, "Reemission Coefficients of Si and SiO_2 Films Deposited Through RF and DC Sputtering," *J. Appl. Phys.*, 38, 4656 (1967); also D.W. Hoffman, "Intrinsic Resputtering—Theory and Experiment," *J. Vac. Sci. Technol.*, A8(5) 3707 (1990)

77. R.L. Sandstrom, W.L. Gallagher, T.R. Dingle, R.H. Koch, R.B. Laibowitz, A.W. Kleinssasser, R.J. Gambino, and M.F. Chisolm, *Appl. Phys. Lett.*, 53, 444 (1986); also A.J. Drehman and M.W. Dumais, "Substrate Bias Effects During RF Sputtering of Y-Ba-Cu-O Films," *J. Mat. Res.*, 5(4) 677 (1990)

78. H.F. Winters and E. Kay, "Gas Incorporation into Sputtered Films," *J. Appl. Phys.*, 38, 3928 (1967); also J.J. Cuomo and R.J. Gambino, "Incorporation of Rare Gases in Sputtered Amorphous Metal Films," *J. Vac. Sci. Technol.*, 14, 152 (1977); also D.M. Mattox and G.J. Kominiak, "Incorporation of Helium in Deposited Gold Films," *J. Vac Sci. Technol.*, 8, 194 (1971)

79. D.J. Ball, *J. Appl. Phys.*, 43, 3047 (1972)

80. J.L. Vossen and J.J. O'Neill, Jr., "DC Sputtering with RF-Induced Bias," *RCA Rev.*, 29, 566 (1968)

81. F. Vratny, B.H. Vromen, and A.J. Harendza-Harinxma, "Anodic Tantalum Oxide Dielectrics Prepared from Body-Centered-Cubic Tantalum and Beta-Tantalum Films," *Electrochem. Technol.*, 5, 283 (1967)

82. "Bias Sputtering of Molybdenum Films," R. Glang, R.A. Holmwood, and P.C. Furois, p. 643 in *Transactions of the 3rd International Vacuum Congress* (1965)

83. E. Kay, "Magnetic Effects on an Abnormal Truncated Glow Discharge and Their Relation to Sputtered Thin-Film Growth," *J. Appl. Phys.*, 34 (4, Part 1) 760 (1963); also K. Wasa and S. Hayakawa, "Efficient Sputtering in a Cold-Cathode Discharge in Magnetron Geometry," *Proceedings of the IEEE*, 55, 2179 (1967); also S.D. Gill and E. Kay, "Efficient Low Pressure Sputtering in a Large Inverted Magnetron Suitable for Film Synthesis," *Rev. Sci. Instrum.*, 36, 277 (1965); also K. Wasa and S. Hayakawa, "Low Pressure Sputtering System of the Magnetron Type," *Rev. Sci. Instrum.*, 40(5) 693 (1969); also Kiyotaka Wasa and Shigeru Hayakawa, "Method of Producing Thin Films by Sputtering," U.S. Patent #3,528,902 (Sept. 15, 1970)

84. W. Knauer, "Ionic Vacuum Pump," (filed Sept. 10, 1962), U.S. Patent #3,216,652 (Nov. 9, 1965)

85. "Alternative Ion Pump Configurations Derived From a More Through Understanding of the Penning Discharge," W. Knauer and E.R. Stack, p. 180 in *Proceedings of the 10th National Vacuum Symposium*, American Vacuum Society (1963)

86. P.J. Clark(e), "Sputtering Apparatus" (filed Nov. 1968), U.S. Patent #3,616,450 (Oct. 26, 1971); (NOTE: on Clarke's patent the name is P.J. Clark which is in error and has been the source of some erroneous references); also D.B. Fraser, "The Sputter (gun) and S-Gun Magnetrons," Ch II-2 in *Thin Film Processes*, edited by John L. Vossen and Werner Kern, Academic Press (1979)

87. J.R. Mullaly, "A Crossed-Field Discharge Device for High Rate Sputtering," RFP-1310, USAEC contract AT929-1-1106 (Nov. 13, 1969); also *R&D Mag.*, p. 40 (Feb. 1971); also (E-11)

88. J.F. Corbani, "Cathode Sputtering Apparatus," (filed July 1973), U.S. Patent #3,878,085 (April 15, 1975)

89. J.S. Chapin, "Sputtering Process and Apparatus," (filed Jan. 1974), U.S. Patent #4,166,018 (Aug. 28, 1979); also *R&D Mag.*, 25 (1) 37 (1974); also "Planar Magnetron Sputtering," R.K. Waits, Ch. II-4 in *Thin Film Processes*, edited by John L. Vossen and Werner Kern, Academic Press (1979); also "Magnetron Sputtering," A.S. Penfold, Sec. A3.2 in *Hand-*

book of Thin Film Process Technology, edited by David A. Glocker and S. Ismat Shah, IOP Publishing (1994)

90. "An Energy Efficient Window System: Final Report," Day Charoudi, Suntek Research Assoc., p. 49 in LBL-9307, UC-95d: EEb-W-79-10, Contract #W-7405-ENG-48 (August 1977)

91. W.W. Carson, "Sputter Gas Pressure and dc Substrate Bias Effects on Thick RF-Diode Sputtered Films of Ti Oxycarbides," J. Vac. Sci. Technol., 12, 845 (1975); also "Recent Advances in Ion Plating," D.M. Mattox, p. 443, Proceedings of the Sixth International Vacuum Congress, (Japan J. Appl. Phys. Suppl. 2, Pt. 1), Kyoto (1974)

92. B. Windows and N. Savvides, "Unbalanced Magnetrons as Sources of High Ion Fluxes," J. Vac. Sci. Technol., A4(3) 453 (1986); also J. Vac. Sci. Technol., A(2) 196 (1986); also N. Savvides and B. Windows, J. Vac. Sci. Technol., A4(3) 504 (1986); also "Unbalanced Magnetron Sputtering," S.L. Rohde, p. 235 in Plasma Sources for Thin Film Deposition and Etching, Physics of Thin Films, Vol. 18, edited by Maurice H. Francombe and John L. Vossen, Academic Press (1994)

93. D.G. Teer, Surf. Coat Technol., 36, 901 (1988); also D.G. Teer, Surf. Coat. Technol., 39/40, 565 (1989); also D.G. Teer, "Magnetron Sputter Ion Plating," U.S. Patent #5,556,519 (Aug. 17,1996); also S. Kadleç, J. Müsil, and W.-D. Münz, "Sputtering Systems With Magnetically Enhanced Ionization for Ion Plating of TiN Films," J. Vac. Sci. Technol., A8(3) 1318 (1990); also Y. Arnal, J. Pelletier, C. Pomot, B. Petit, and A. Durandet, Appl. Phys. Lett., 45, 132 (1984)

94. Peter D. Davidse, Joseph S. Logan, and Fred S. Maddocks, "Method for Sputtering a Film on an Irregular Surface," U.S. Patent #3,755,123 (Aug. 28, 1973)

95. F. Vratny "Deposition of Tantalum and Tantalum Oxide by Superimposing RF and DC Sputtering," J. Electrochem. Soc., 114(5) 505 (1967)

96. R.L. Cormia "Method for Coating a Substrate," U.S. Patent #4,046,659 (Sept. 6, 1977); also Robert Cormia, "Oral History Interview," Society of Vacuum Coaters (2002), available from SVC; also www.svc.org

97. H.E. McKelvey, "Magnetron Cathode Sputtering Apparatus," U.S. Patent #4,356,073 (Oct. 26, 1982)

98. Russ Hill, private communication

99. Fazle S. Quazi, "Method and Apparatus for Sputtering a Dielectric Target or for Reactive Sputtering," (filed Feb. 1986), U.S. Patent #4,693,805 (Sept. 15, 1987)

100. E.M. Honig, "Generation of a 75-MW 5-kHz Pulse Train from an Inductive Energy Store," IEEE Trans. Plasma Sci., PS-12(1) 24 (1984); also G.A. Farrall, "Arc Extinction Phenomena in Vacuum," J. Appl. Phys., 42(8) 3084 (1971)

101. P. Frach, U. Heisig, Chr. Gottfried, and H. Walde, "Aspects and Results of Long-Term Stable Deposition of Al_2O_3 With High Rate Pulsed Reactive Magnetron Sputtering," Surf. Coat. Technol., 59, 177 (1993)

102. G. Este and W.D. Westwood "A Quasi-direct-current Sputtering Technique for the Deposition of Dielectrics at Enhanced Rates," J. Vac. Sci. Technol., A6(3) 1845 (1988)

103. D.A. Glocker, private communication; also D.A. Glocker, "Influence of the Plasma on Substrate Heating During Low Frequency Reactive Sputtering of AlN," J. Vac. Sci. Technol., A11(6) 2989 (1993)

104. M. Scherer, J. Schmitt, R. Latz, and M. Schanz, "Reactive Alternating Current Magnetron Sputtering of Dielectric Layers," J. Vac. Sci. Technol., A10(4) 1772 (1992)

105. "AC Reactive Sputtering with Inverted Cylindrical Magnetrons," D.A. Glocker, V.W. Lindberg, and A.R. Woodard, p. 81 in Proceedings of the 43rd Annual Technical Conference, Society of Vacuum Coaters (2000)

106. W.-D. Münz, Surf. Coat. Technol., 48, 81 (1991); also W.-D. Münz and F.J.M. Hauser, Surf. Coat. Technol., 49, 161 (1991)

106a. J.J. Cuomo and S.M. Rossnagel, "Hollow-Cathode-Enhanced Magnetron Sputtering," J. Vac. Sci. Technol., A4, 393 (1986); also J.M. Schneider, A.A. Voevodin, C. Rebholz, and A. Matthews, "Microstructural and Morphological Effects on the Tribological Properties of Electron Enhanced Magnetron Sputtered Hard Coatings," J. Vac. Sci. Technol., A13(4) 2189 (1995)

107. Douglas S. Schatz and Richard A. Scholl, "Continuous Deposition of Insulating Material Using Multiple Anodes Alternating Between Positive and Negative Voltages," U.S. Patent #5,897,753 (Apr. 27, 1999); also "Reactive Sputtering Using a Dual-Anode Magnetron System," A. Belkind, Z. Zhao, D. Carter, G. McDonough, G. Roche, and R. Scholl, p. 130 in Proceedings of the 44th Annual Technical Conference, Society of Vacuum Coaters (2001)

108. "Flat Plate Magnetron Sputtering Device," Chikara Hatashi, Komiya Muneharu, and Kusumoto Toshio, Japanese application #JP1985000154062 (July 15, 1985); also S.M. Rossnagel, D. Mikalsen, H. Kinoshita, and J.J. Cuomo, "Collimated Magnetron

Sputter Deposition," *J. Vac. Sci. Technol.*, A9(2) 261 (1991); also Steven Hurwitt, "Extended Lifetime Collimator," U.S. Patent #5,223,108 (1993)

108a. N. Motegi et al, "Long-Throw Low-Pressure Sputtering Technology for Very Large-Scale Integrated Devices," *J. Vac. Sci. Technol.*, B13(4) 1906 (1995)

109. Th. Jung and A. Westphal, "High Rate Deposition of Alumina Films by Reactive Gas Flow Sputtering," *Surf Coat Technol.*, 59, 171 (1993)

110. S.M. Rossnagel and J. Hopwood, "Metal Deposition from Ionized Magnetron Sputtering," *J. Vac. Sci. Technol.*, B12(1) 449 (1994)

111. R.C. Krutenat and W.R. Gesick, "Vapor Deposition by Liquid Phase Sputtering," *J. Vac. Sci. Technol.*, 7(6) S40 (1970)

112. N. Hosokawa, T. Tsukada, and H. Kitahara, *Proceedings of the 8th International Vacuum Congress*, Cannes, France, *LeVide Suppl.*, 201, 11 (1980); also R. Kukla, T. Krug, R. Ludwig, and K. Wilmes, *Vacuum*, 41, 1968 (1990); also W.M. Posadowski, *Surf. Coat. Technol.*, 49, 290 (1991); also "Self-Sputtering With DC Magnetron Source: Target Material Considerations," Z.J. Radzimski and W.M. Posadowski, p. 389 in *Proceedings of the 37th Annual Technical Conference*, Society of Vacuum Coaters (1994)

113. K.L. Chopra and M.R. Randlett, *Rev. Sci. Instrum.*, 38, 1147 (1967); also P.H. Schmidt, R.N. Castellano, and E.G. Spenser, *Solid State Technol.*, 15, 27 (1972); also "Ion Beam Deposition," J.M.E. Harper, Sec. II-5 in *Thin Film Processes*, edited by John L. Vossen and Werner Kern, Academic Press (1979); also W.D. Westwood and S.J. Ingrey, "Fabrication of Optical Wave-guides by Ion Beam Sputtering," *J. Vac. Sci. Technol.*, 13, 104 (1976); also *Ion Beam Etching, Sputtering and Plating*, edited by C. Weissmantel and G. Gautherin, Elsevier (1978)

114. H.R. Kaufman, J.J. Cuomo, and J.M.E. Harper, "Technology and Application of Broad-Beam Ion Sources Used in Sputtering: Part 1—Ion Source Technology," *J. Vac. Sci. Technol.*, 21(3) 725 (1982)

115. R.P. Netterfield, W.G. Sainty, P.J. Martin, and S.H. Sie, "Properties of CeO_2 Thin Films Prepared by Oxygen-Ion-Assisted Deposition," *Appl. Opt.*, 24, 2267 (1985)

116. David T. Wei and Anthony W. Louderback, "Method For Fabricating Multi-Layer Optical Films," Patent #4,142,958 (March 6, 1979)

117. D.M. Mattox and G.J. Kominiak, "Structure Modification by Ion Bombardment During Deposition," *J. Vac. Sci. Technol.*, 9, 528 (1972); also R.D. Bland, G.J. Kominiak, and D.M. Mattox, "Effect of Ion Bombardment During Deposition of Thick Metal and Ceramic Deposits," *J. Vac Sci. Technol.*, 11, 671 (1974); also K.H. Müller, "Monte Carlo Calculations for Structural Modification in Ion-Assisted Thin Film Deposition," *J. Vac Sci Technol.*, A4(2) 184 (1986)

118. B.A. Movchan and A.V. Demchishin, "Study of the Structure and Properties of Thick Vacuum Condensates of Nickel, Titanium, Tungsten, Aluminum Oxide and Zirconium Oxide," *Phys. Met. Metalogr.*, (translation from the Russian), 28, 83 (1969); also A. Van der Drift, "Evolutionary Selection: A Principle Governing Growth Orientation in Vapour-Deposited Layers," *Philips Res. Report*, 22, 267 (1967); also A.G. Dirks and H.J. Leamy, "Columnar Microstructure in Vapor Deposited Thin Films," *Thin Solid Films,* 47, 219 (1977)

119. J.A. Thornton, "High Rate Thick Film Growth," *Ann. Rev. Mater. Sci.*, 7, 239 (1977)

120. R. Messier, A.P. Giri, and R.A. Roy, "Revised Structure Zone Model for Thin Film Physical Structure," *J. Vac Sci. Technol.*, A2, 500 (1984)

121. J.W. Patten, "The Influence of Surface Topography and Angle of Adatom Incidence on Growth Structure in Sputtered Chromium," *Thin Solid Films*, 63, 121 (1979); also P. Bai, J.F. McDonald, and T.M. Lu, "Effects of Surface Roughness on the Columnar Growth of Cu Films," *J. Vac. Sci. Technol.*, A9(4) 2113 (1991)

122. N.O. Young and J. Kowal, *Nature,* 183, 104 (1959); also K. Robbie, M.J. Brett, and A. Lakhtakia, *J. Vac. Sci. Technol.*, A13, 2991 (1995); also K. Robbie and M.J. Brett, "Sculpted Thin Films and Glancing Angle Deposition: Growth Kinetics and Applications," *J. Vac. Sci. Technol.*, 15(3) 1460 (1997); also K. Robbie and M.J. Brett, U.S. Patent #5,866,204 (1999); also R. Messier, T. Gehrke, C. Frankel, V.C. Venugopal, W. Otaño, and L. Lakhtakia, *J. Vac. Sci. Technol.*, A15, 2148 (1997); also R. Messier, A. Lakhtakia, V.C. Venugopal, and P. Sunal, "Sculptured Thin Films: Engineered Nanostructural Materials," *Vac. Technol. Coat.*, 2(10) 40 (2001); also M. Malac, R. Egerton, and M. Brett, "Thin Films Deposited at Glancing Incidence and Their Applications," *Vac. Technol. Coat.*, 2(7) 48 (2001)

123. J.A. Thornton and D.W. Hoffman, *J. Vac. Sci. Technol.*, 14, 164 (1977); also D.W. Hoffman and J.A. Thornton, "Effects of Substrate Orientation and Rotation on the Internal Stresses in Sputtered Metal Films," *J. Vac. Sci. Technol.*, 16, 134 (1979)

124. D.M. Mattox, R.E. Cuthrell, C.R. Peeples, and P.L. Dreike, "Preparation of Thick Stress-Free Mo Films for a Resistively Heated Ion Source," *Surf. Coat. Technol.*, 36, 117 (1988); also "Residual Stress, Fracture and Adhesion in Sputter Deposited Mo-

lybdenum Films," D.M. Mattox and R.E. Cuthrell, p. 141 in *Adhesion in Solids*, edited by D.M. Mattox, J.E.E Baglin, R.E. Gottchall, and C.D. Batich, Vol. 119, *MRS Symposium Proceedings* (1988)

125. W.D. Sproul and James R. Tomashek, "Rapid Rate Reactive Sputtering of a Group IVB Metal," U.S. Patent #4,428,811 (Jan. 31, 1984)

126. J.E. Greene and F. Sequeda-Osorio, "Glow Discharge Spectroscopy for Monitoring Sputter Deposited Film Thickness," *J. Vac. Sci. Technol.*, 10(6) 1144 (1973); also S. Schiller, U. Heisig, K. Steinfelder, J. Strümpfel, R. Voigt, R. Fendler, and G. Teschner, "On the Investigation of d.c. Plasmatron Discharges by Optical Emission Spectroscopy," *Thin Solid Films*, 96, 235 (1982): also S. Schulz and J. Strümpfel, "Plasma Emission Monitoring," *Vac. Technol. Coat.*, 2(8) 42 (2001)

127. B.J. Curtis, "Optical End-Point Detection for Plasma Etching of Aluminum," *Solid State Technol.*, 23(4) 129 (1980); also J.A. Thornton, "Diagnostic Methods for Sputtering Plasmas," *J. Vac. Sci. Technol.*, 15(2) 188 (1978); also R.W. Dreyfus, J.M. Jasinski, R.E. Walkup, and G.S. Selwyn, "Optical Analysis of Low Pressure Plasmas," *Pure and Applied Chemistry*, 57(9) 1265 (1985)

128. J. Stark, *Z. Elektrochem.*, 14, 752 (1908); also *Z. Elektrochem.*, 15, 509 (1909)

129. "The Nature of Physical Sputtering," G.K. Wehner, Chapter 3, p. 3-33, in *Handbook of Thin Film Technology*, edited by Leon I. Maissel and Reinhard Glang, McGraw-Hill Book Co. (1970)

130. *Rays of Positive Electricity*, J.J. Thompson, p. 103, Longmans, Green & Co., London (1913)

131. "Evaporation and Sputtering," Ch. IV in *Procedures in Experimental Physics*, J. Strong, Prentice-Hall (1938)

132. "Vacuum Evaporation," Reinhard Glang, Chapter 1 in *Thin Film Technology*, edited by Leon I. Maissel and Reinhard Glang, McGraw-Hill Book Co. (1970)

133. "Evaporation," Rointan F. Bunshah, Ch. 4 in *Deposition Technologies for Films and Coatings: Development and Applications*, edited by Rointan F. Bunshah et al, Noyes Publications (1982), and second edition William Andrew Publishing/Noyes Publications (1994)

134. "Vacuum Evaporation and Vacuum Deposition," Ch. 5, Sec. 5.2.1 in *Handbook of Physical Vapor Deposition (PVD) Processing*, Donald M. Mattox, William Andrew Publishing/Noyes Publications (1998)

135. H. Hertz, *Ann. Physik (Leipzig)*, 17, 177 (1882); also *Wied. Ann.*, 17, 193 (1882)

136. S. Stefan, *Wien. Ber.*, 68, 385 (1873); also S. Stefan, *Wien. Ber.*, 98, 1418 (1889)

137. M. Knudsen, *Ann. Physik*, 47, 697 (1915)

138. R.E. Honig, "Vapor Pressure Data for the More Common Elements," *RCA Review*, 18, 195 (1957); also R.E. Honig and H.O. Hook, "Vapor Pressure for Some Common Gases," *RCA Review*, 21, 360 (1960); also R.E. Honig, "Vapor Pressure Data for the Solid and Liquid Elements," *RCA Review*, 23, 567 (1962); also R.E. Honig and D.A. Kramer, "Vapor Pressure Data for the Solid and Liquid Element," *RCA Review*, 30, 285 (1969)

139. R. Nahrwold, *Ann. Physik*, 31, 467 (1887); also A. Kundt, *Ann. Physik*, 34, 469 (1888)

140. J. Stuhlmann, *J. Am. Opt. Soc.*, 1, (2) 78 (1917)

141. F. Soddy, *Proc. Roy. Soc. (London)*, 78, 429 (1907); also *Proc. Roy. Soc. (London)*, A80, 92 (1908); also I. Langmuir, *J. Am. Chem. Soc.*, 35, 931 (1913)

142. R. von Pohl and P. Pringsheim, "Über die Herstellung von Metallspiegeln durch Distillation im Vakuum," *Verhandl. Deut. Physik. Ges.*, 14, 506 (1912); NOTE: see endnote E-14

143. I. Langmuir, *Phys. Rev.*, 2, 329 (1913)

144. R. Ritschl, *Zeits. F. Physik*, 69, 578 (1931); NOTE: see endnote E-15

145. C.H. Cartwright and J. Strong, *Rev. Sci. Instrum.*, 2, 189 (1931)

146. J. Strong, *Phys. Rev.*, 43, 498 (1933)

147. J. Strong, *Phys. Rev.*, 39, 1012 (1932); also R.C. Williams, *Phys. Rev.*, 41, 255 (1932)

148. J. Strong, *Astrophysical Journal*, 83(5) 401 (1936)

149. *The Perfect Machine: The Building of the Palomar Telescope*, Ronald Florence, pp 382-386, HarperCollins (1994)

150. "The Design and Operation of Large Telescope Mirror Aluminizers," J.A.F. Trueman, p. 32 in *Proceedings of the 22nd Annual Technical Conference*, Society of Vacuum Coaters (1979)

151. F. Adams "Vacuum Metallizing in the Lamp Industry," p. 48 in *Proceedings of the 23rd Annual Technical Conference*, Society of Vacuum Coaters (1980)

152. A.H. Pfund, "Highly Reflecting Films of Zinc Sulfide," *J. Opt. Soc. Am.*, 24, 99 (1934); also A. Macleod, "The Early Days of Optical Coatings," *J. Opt. A: Pure Appl. Opt.*, 1, 779 (1999)

153. G. Bauer, *Ann. Phys. (Leipzig)*, 19, 434 (1934)

154. A. Smakula, German Patent #DRP 685767 (filed 5 Oct. 1935) (May 1940)

155. J. Strong, "On a Method of Decreasing the Reflection from Non-metallic Substrate," *J. Optical Soc. Am.*, 36, 73 (1936)

156. C.H. Cartwright and A.F. Turner, *Phys. Rev.*, 55, 675 (1939)

157. P. Baumeister, "The First Calculations of the Quarterwave Stack," *Proceedings of the 46th Annual Technical Conference*, Society of Vacuum Coaters (2003)—to be published

158. Collin Alexander, private communication

159. Richard A. Denton, "The Manufacture of Military Optics at the Frankfort Arsenal During WWII," *Optics News* 15, 24 (1989)

160. D.A. Lyon, U.S. Patent #2,398,382 (Nov. 1942)

161. The text of "Application of Metallic Fluoride Reflection Reducing Films to Optical Elements" may be found in the History section of the Society of Vacuum Coaters web site: www.svc.org

162. "Optical Thin Film Technology—Past, Present and Future," W.P. Strickland, p. 221 in *Proceedings of the 33rd Annual Technical Conference*, Society of Vacuum Coaters (1990) [information from "Report on German Vacuum Evaporation Methods of Producing First Surface Mirrors, Semi-transparent Mirrors and Non-Reflecting Films," Combined Intelligence Objectives Sub-Committee, Intell. Div., TSFET, Report #H-2 (1945)]; also J. Strong, *Rev. Sci. Instrum.*, 6, 97 (1935)

163. Combined Intelligence Objectives Sub-committee File #XXVII-44, Item #1, "Manufacture of Metallized Paper Capacitor Units," (Germany); also [33]

164. R. Bosch, British Patent #510,642 (Aug. 1937)

165. I. Langmuir, *Proc. Nat. Acad. Sci., Wash.*, 3, 141 (1917)

166. "Vacuum Cadmium Plating," Military Specification MIL-C-8837—ASG, 3 (June 1958)

167. M. von Pirani, "Production of Homogeneous Bodies From Tantalum or Other Materials," U.S. Patent #848,600 (Mar. 26, 1907)

168. H.M. O'Brian and H.W.B. Skinner, *Phys. Rev.*, 44, 602 (1933); also H.M. O'Brian, *Rev. Sci. Instrum.*, 5, 125 (1934)

169. L. Holland, British Patent #754,102 (1951); also p. 135 in *Vacuum Deposition of Thin Films*, L. Holland, Chapman Hall (1957)

170. *Theory and Design of Electron Beams*, J.R. Pierce, D. Van Nostrand (1949)

171. S. Schiller, and G. Jäsch, "Deposition by Electron Beam Evaporation with Rates up to 50µm/s," *Thin Solid Films*, 54, 9 (1978)

172. B.A. Unvala and G.R. Booker, *Phil. Mag.*, 9, 691 (1964)

173. C.W. Hanks, "Apparatus for Producing and Directing an Electron Beam," U.S. Patent #3,535,428 (Oct. 20, 1970)

174. "High Rate Horizontally Emitting Electron Beam Vapor Source," H.R. Smith, p. 49 in *Proceedings of the 21st Annual Technical Conference*, Society of Vacuum Coaters (1978)

175. F.L. Schuermeyer, W.R. Chase, and E.L. King, "Self-Induced Sputtering During Electron-Beam Evaporation of Ta," *J. Appl. Phys.*, 42, 5856 (1971); also F.L. Schuermeyer, W.R. Chase, and E.L. King, "Ion Effects During E-Beam Deposition of Metals," *J. Vac. Sci. Technol.*, 9, 330 (1972)

176. "Development of Processing Parameters and Electron-Beam Techniques for Ion Plating," D.L. Chambers and D.C. Carmichael, p. 13 in *Proceedings of the 14th Annual Technical Conference*, Society of Vacuum Coaters (1971)

177. J.R. Morley and H. Smith, "High Rate Ion Production for Vacuum Deposition," *J. Vac. Sci. Technol.*, 9, 1377 (1972); also S. Komiya and K. Tsuruoka, *J. Vac. Sci. Technol.*, 12, 589 (1975); also Y.S. Kuo, R.F. Bunshah, and D. Okrent, "Hot Hollow Cathode and its Application in Vacuum Coating: A Concise Review," *J. Vac. Sci. Technol.*, A4(3), 397 (1986)

178. R.G. Picard and J.E. Joy, *Electronics*, 24, 126 (April 1951)

179. I. Ames, L.H. Kaplan, and P.A. Roland, "Crucible Type Evaporation Source for Aluminum," *Rev. Sci. Instrum.*, 37, 1737 (1966)

180. T. Hibi, *Rev. Sci. Instrum.*, 23, 383 (1952)

181. "The Significance of Impact Velocity of Vacuum-Deposited Atoms for the Structure of Thin Films," H. Fuchs and H. Gleiter, p. 4 in *Thin Films: The Relationship of Structure to Properties*, edited by C.R. Aita and K.S. SreeHarsha, *MRS Proceedings*, Vol. 47 (1985)

182. R.J. Ney, "Nozzle Beam Evaporation Source," *J. Vac. Sci. Technol.*, A1(1) 55 (1983)

183. J.J. Schmitt, "Method and Apparatus for the Deposition of Solid Films of Material from a Jet Stream Entraining the Gaseous Phase of Said Material," U.S. Patent #4,788,082 (Nov. 1988); also J.F. Groves, G. Mattausch, H. Morgner, D.D. Hass, and H.N.G. Wadley, "Directed Vapour Deposition," *Surf. Eng.*, 6(6) 461 (2000)

184. *Scientific Foundations of Vacuum Techniques*, S. Dushman, 1st edition, John Wiley (1949); also A.D. Romig, Jr., "A Time Dependent Regular Solution Model for the Thermal Evaporation of an Al-Mg Alloy," *J. Appl. Phys.*, 62, 503 (1987)

185. T. Santala and M. Adams, *J. Vac. Sci. Technol.*, 7, s22 (1970); also H.R. Barker and R.J. Hill, "The Deposition of Multicomponent Phases in Ion Plating," *J. Vac Sci. Technol.*, 9(6) 1395 (1972)

186. L. Harris and B.M. Siegel, "A Method for the

Evaporation of Alloys," *J. Appl. Phys.*, 19, 739 (1948); also "Flash Evaporation," J.L. Richards, p. 71 in *The Use of Thin Films in Physical Investigations*, edited by J.C. Anderson, Academic Press (1966)

187. "Methods of Continuous High Vacuum Strip Processing," H.F. Smith, Jr., and C.d'A. Hunt, *Transactions of the Vacuum Metallurgy Conference*, AVS Publications (1964): also R.F. Bunshah and R.S. Juntz, p. 200 in *Transactions of the Vacuum Metallurgy Conference*, AVS Publications (1965)

188. A.H.F. Muggleton, "Deposition Techniques for Preparation of Thin Film Nuclear Targets: Invited Review," *Vacuum*, 37, 785 (1987)

189. "Silicon Monoxide Evaporation Methods," C.E. Drumheller, p. 306 in *Transactions of the 7th AVS Symposium*, Pergamon Press (1960); also W.C. Vergara, H.M. Greenhouse, and N.C. Nicholas, *Rev. Sci. Instrum.*, 34, 520 (1963)

190. loc. cit. [152]

191. G. Hass, *J. Am. Chem. Soc.*, 33, 353 (1950); also G. Hass and C.D. Salzberg, *J. Opt. Soc. Am.*, 44, 181 (1954)

192. M. Aüwarter, Austrian Patent #192,650 (1952); also D.S. Brinsmaid, G.J. Koch, W.J. Keenan, and W.F. Parson, U.S. Patent #2,784,115 (filed May 1953) (1957); also E. Cremer, T. Kraus, and E. Ritter, *Z. Elektrochem.*, 62, 939 (1958); also E. Ritter, "Deposition of Oxide Films by Reactive Evaporation," E. Ritter, *J. Vac. Sci. Technol.*, 3(4) 225 (1966)

193. M. Aüwarter, U.S. Patent #2,920,002 (1960)

194. J.T. Cox, G. Hass, and I.B. Ramsey, *J. Phys. Radium*, 25, 250 (1964)

195. R. Nimmagadda, A.C. Raghuram, and R.F. Bunshah, "Activated Reactive Evaporation Process for High Rate Deposition of Compounds," *J. Vac. Sci. Technol.*, 9(6) 1406 (1972); also "Activated Reactive Evaporation (ARE)," R.F. Bunshah, *Handbook of Deposition Technologies for Films and Coatings*, 2nd edition, p. 134, edited by R.F. Bunshah et al, William Andrew Publishing/Noyes Publications (1994); also K. Nakmura, K. Inagawa, K. Tsuruoka, and S. Komiya, "Applications of Wear-Resistant Thick Films Formed by Physical Vapor Deposition Processes," *Thin Solid Films*, 40, 155 (1977)

196. "Gas Evaporation and Ultrafine Particles," sec. 5.12, p. 301 in *Handbook of Physical Vapor Deposition (PVD) Processing*, Donald M. Mattox, William Andrew Publishing/Noyes Publications (1998); also C. Hayashi, "Ultrafine Particles," *Physics Today*, 40, 44 (1987); also G.D. Stein, "Cluster Beam Sources: Predictions and Limitations of the Nucleation Theory," *Surf. Sci.*, 156, 44 (1985); also J.K.G.

Panitz, D.M. Mattox, and M.J. Carr, "Salt Smoke: The Formation of Submicron Sized RbCl Particles by Thermal Evaporation in 0.5-1200 Torr of Argon and Helium," *J. Vac. Sci. Technol.*, A6 (6), 3105 (1988)

197. A.H. Pfund, "The Optical Properties of Metallic and Crystalline Powders," *J. Opt. Soc. Am.*, 23, 375 (1933): also D.M. Mattox and G.J. Kominiak, "Deposition of Semiconductor Films with High Solar Absorbtivity," *J. Vac. Sci. Technol.*, 12(1) 182 (1975)

198. T. Takagi, I. Yamada, K. Yanagawa, M. Kunori, and S. Kobiyama, "Vaporized-Metal Cluster Ion Source for Ion Plating," *Jpn. J. Appl. Phys., Suppl. 2, Pt. 1*, p. 427 (1974); also *Ionized Cluster Beam Deposition and Epitaxy*, T. Takagi, William Andrew Publishing/Noyes Publications (1988)

199. H. Haberland, M. Karrasis, M. Mall, and Y. Thurner, "Thin Films From Energetic Cluster Impact: A Feasibility Study," *J. Vac. Sci. Technol.*, A10 (5) 3266 (1992); also H . Haberland, M. Mall, M. Mosler, Y. Qiang, T. Reiners, and Y. Thurner, "Filling of Micron-Sized Contact Holes With Copper by Energetic Cluster Impact," *J. Vac. Sci. Technol.*, A12(5) 2925 (1994)

200. H.M. Smith and A.F. Turner, "Vacuum Deposited Thin Films Using a Ruby Laser," *Appl. Optics*, 4, 147 (1965); also J.T. Cheung and H. Sankur, "Growth of Thin Films by Laser-Induced Evaporation," *Crit. Rev. Solid State Mat. Sci.*, 15, 63 (1988)

201. *Pulsed Laser Deposition,* edited by D.B. Chrisey and G.K. Hubler, John Wiley (1994)

202. D. Dijkkamp, T. Venkatesan, X.D. Wu, S.A. Shaheen, N. Jisrawi, Y.H. Min-Lee, W.L. McLean, and M. Croft, *Appl. Phys. Lett.*, 51, 619 (1987); also J.T. Cheung and J. Madden, *J. Vac. Sci. Technol.*, B5, 705 (1987)

203. D.W. Pashley, *Adv. Phys.*, 5, 173 (1956)

204. K.Z. Günther, *Z. Naturforsch*, 13a, 1081 (1958)

205. J.E. Davey and T. Pankey, *J. Appl. Phys.*, 39, 1941 (1968)

206. A.Y. Cho and J.R. Arthur, *Progress in Solid State Chemistry*, edited by G. Somorjai and J. McCaldin, Pergamon Press (1975)

207. J.R. Knight, D. Effer, and P.R. Evans, *Solid State Electronics*, 8, 178 (1965)

208. M. Banning, "Practical Methods of Making and Using Multilayer Filters," *J. Opt. Soc. Am.*, 37, 792 (1945)

209. W. Steckelmacher, J.M. Parisot, L. Holland, and T. Putner, *Vacuum*, 9, 171 (1959)

210. G. Sauerbrey, *Phys. Verhandl.*, 8, 113 (1957); also

Z. Physik, 155, 206 (1959); also M.P. Lotis, *J. Phys. Radium*, 20, 25 (1959)

211. D.I. Perry, "Low Loss Multilayer Dielectric Mirrors," *Appl. Optics*, 4, 987 (1965); also *Laser Induced Damage in Optical Materials 1979*, edited by Harold E. Bennett, Alexander J. Glass, Arthur H. Guenther, and Brian E. Newman, *National Bureau of Standards Special Publication #568* (1980)

212. K.H. Behrndt, "Films of Uniform Thickness Obtained from a Point Source," *Transactions of the Ninth Vacuum Symposium*, American Vacuum Society, p. 111 (1962)

213. D.A. Buck and K.R. Shoulders, "An Approach to Microminiature Systems," pp. 55-59, *Proceedings of the Eastern Joint Computer Conference*, American Institute of Electrical Engineers (1958)

214. A. Yializis, G.L. Powers, and D.G. Shaw, "A New High Temperature Multilayer Capacitor with Acrylate Dielectrics," *IEEE Tranactions on Components, Hybrids, and Manufacturing Technology*, Vol. 13, #4 (Dec. 1990); also D.G. Shaw and M.G. Langlois, "Use of Vapor Deposited Acrylate Coatings to Improve the Barrier Properties of Metallized Film," p. 240 in *Proceedings of the 37th Annual Technical Conference*, Society of Vacuum Coaters (1994); also P.M. Martin, J.D. Affinito, M.E. Moss, C.A. Coronado, W.D. Bennett, and D.C. Stewart, "Multilayer Coatings on Flexible Substrates," p. 163 in *Proceedings of the 38th Annual Technical Conference*, Society of Vacuum Coaters (1995); also J. Affinito, "Polymer Film Deposition by a New Vacuum Process," p. 425 in *Proceedings of the 45th Annual Technical Conference*, Society of Vacuum Coaters (2002)

215. R. Olsen, "The Application of Thin Vacuum-Deposited Poly-para-xylene to Provide Corrosion Protection for Thin, Porous Inorganic Coatings," p. 317 in *Proceedings of the 34th Annual Technical Conference*, Society of Vacuum Coaters (1991)

216. *Handbook of Vacuum Arc Science and Technology: Fundamentals and Applications*, edited by Raymond L. Boxman, Philip J. Martin, and David M. Sanders, William Andrew Publishing/Noyes Publications (1995)

217. "Cathodic Arc Deposition," P.J. Martin, Sec. A1.3 in *Handbook of Thin Film Process Technology*, edited by David A. Glocker and S. Ismat Shah, Institute of Physics (IOP) Publishing (1995)

218 "Vacuum Arc-Based Processing," D. Sanders, Sec. V-18 in *Handbook of Plasma Processing Technology*, William Andrew Publishing/Noyes Publications (1990)

219. "The Past, Present and Future of Controlled Atmospheric Arcs," W.E. Kuhn, p.1 in *Arcs in Inert Atmospheres and Vacuum*, edited by W.E. Kuhn, John Wiley (1956); also C.A. Doremus, *Trans. Electrochem. Soc.*, 13, 347 (1908); *Electric Arcs: Experiments Upon Arcs Between Different Electrodes in Various Environments and Their Explanation*, Clement D. Child, Van Nostrand (1913); also H. Wroe, "Stabilization of Low Pressure D.C. Arc Discharges," U.S. Patent # 2,972,695 (Feb. 21, 1961); also F.J. Zanner and L.A. Bertram, "Behavior of Sustained High-Current Arcs on Molten Alloy Electrodes During Vacuum Consumable Arc Remelting," *IEEE Transactions on Plasma Science*, PS-11 (3) 223 (1983)

220. M. Faraday, "Experimental Relations of Gold (and Other Metals) to Light," (A Bakerian Lecture) *Phil. Trans.*, 147, 145 (1857) and reprinted in *Exp. Res. Chem. Phys.*, p. 391 (1857); also *Michael Faraday*, L. Pearce Williams, p. 474, Basic Books, Inc. (1965)

221. E. Moll and H. Daxinger, U.S. Patent #4,197,175 (1980); also R. Buhl, E. Moll, and H. Daxinger, "Method and Apparatus for Evaporating Material Under Vacuum Using Both Arc Discharge and Electron Beam," U.S. Patent #4,448,802 (1984); also H.K. Pulker, "Method of Producing Gold-Color Coatings," U.S. Patent #4,254,159 (Mar. 3, 1981)

222. S. Komiya and K. Tsuroka, "Thermal Input to Substrate During Deposition by Hollow Cathode Discharge," *J. Vac. Sci. Technol.*, 12, 589 (1975); also D.S. Komiya, "Physical Vapor Deposition of Thick Cr and its Carbide and Nitride Films by Hollow Cathode Discharge," *J. Vac. Sci. Technol.*, 13, 520 (1976)

223. *Atom and Ion Sources*, L. Valyi, John Wiley (1977); also J. Druaux and R. Bernas, p. 456 in *Electromagnetically Enriched Isotopes and Mass Spectrometry*, edited by M.L. Smith, Academic Press (1956)

224. D.E. Bradley, *Brit. J. Appl. Phys.*, 5, 65 (1954); also M.D. Blue and G.C. Danielson, "Electrical Properties of Arc-Evaporated Carbon Films," *J. Appl. Phys.*, 28, 583 (1957); also "Production of Self-Supporting Carbon Films," B.J. Massey, p. 922 in *Transactions of the 8th AVS National Symposium*, Pergamon Press (1961)

225. "A New Deposition Technique for Refractory Metal Films," M.S.P. Lucas, C.R. Vail, W.C. Stewart, and H.A. Owens, p. 988 in *Transactions of the 8th AVS National Symposium*, Pergamon Press (1962); also H. Wroe, "The Magnetic Stabilization of Low Pressure D.C. Arcs," *Brit. J. Appl. Phys.*, 9, 488 (1958)

226. M. Kikushi, S. Nagakura, H. Ohmura, and S. Oketani, "Structure of Metal Films Produced by Vacuum-Arc Evaporation Method," *Jpn. J. Appl. Phys.*, 4, 940 (1965)

227. A.R. Wolter, p. 2A-1 in *Proceedings of the 4th Microelectronic Symposium*, IEEE (1965)

228. A.M. Dorodnov, A.N. Kuznetsov, and V.A. Petrosov, "New Anode-Vapor Vacuum Arc With a Permanent Hollow Cathode," *Sov. Tech. Phys. Lett.*, 5(8) 419 (1979); also Y.S. Kuo, R.F. Bunshah, and D. Okrent, "Hot Hollow Cathode and its Application in Vacuum Coating: A Concise Review," *J. Vac. Sci. Technol.*, A4(3) 397 (1986); also H.K. Pulker, "Ion Plating as an Industrial Manufacturing Method," *J. Vac. Sci. Technol.*, A10(4) 1669 (1992)

229. H. Ehrich, "The Anodic Vacuum Arc: Basic Construction and Phenomenology," *J. Vac. Sci. Technol.*, A6(1) 134 (1988); also *Vacuum Technik*, 37(3) 176 (1988)

230. A.A. Snaper, "Arc Deposition Process and Apparatus," U.S. Patent #3,836,451 (1974)

231. Leonid Pavlovich Sablev, Nikolai Petrovich Atamansky, Valentin Nikolaevich Gorbunov, Jury Ivanovich Dolotov, Vadim Nikolsevich Lutseenko, Valentin Mitro Fanovich Lunev, and Valdislav Vasilievich Usov, "Apparatus for Metal Evaporation Coating," U.S. Patent #3,793,179 (filed July 19, 1971) (Feb. 19, 1974)

232. W.M. Mularie, U.S. Patent #4,430,184 (1984)

233. A. Gilmour and D.L. Lockwood, "Pulsed Metallic-Plasma Generator," *Proceedings of the IEEE*, 60, 977 (1972)

234. P. Siemroth and H.-J. Scheibe, "The Method of Laser-Sustained Arc Ignition," *IEEE Trans. Plasma Sci.*, 18, 911 (1990)

235. Gary E. Vergason, "Electric Arc Vapor Deposition Device," U.S. Patent #5,037,522 (filed July 24, 1990) (Aug. 6, 1991); also K.R. Narendrnath and C. Zhao, *J. Adhesion Sci. Technol.*, 6, 1179 (1991)

236. Richard P. Welty, "Apparatus and Method for Coating a Substrate Using Vacuum Arc Evaporation," U.S. Patent #5,269,898 (filed Dec. 1991) (Dec. 24, 1993)

237. J.E. Daalder, "Components of Cathode Erosion in Vacuum Arc," *J. Phys., D. Appl. Phys.*, 9, 2379 (1976); also "Cathode Spots and Vacuum Arcs," *Phys. Stat. Solid.*, 104, 91 (1981); also C.W. Kimblin, "Cathode Spot Erosion and Ionization Phenomena in the Transition from Vacuum to Atmospheric Pressure Arcs," *J. Appl. Phys.*, 45(12) 5235 (1974)

238. I.I. Aksenov, V.A. Belous, and V.G. Padalka, "Apparatus to Rid the Plasma of a Vacuum Arc of Macroparticles," *Instrum Exp. Tech.*, 21, 1416 (1978); also I.I. Aksenov, V.A. Padalka, and A.I. Popov, *Sov. J. Plasma Phys.*, 4, 425 (1978); also I.I. Aksenov, A.N. Belokhvotikov, V.G. Padalka, N.S. Repalov, and V.M. Khoroshikh, *Plasma Physics Control Fusion*, 28, 761 (1986); also P.J. Mar-

tin, R.P. Netterfield, and T.J. Kinder, "Ion-Beam-Deposited Films Produced by Filtered Arc Evaporation," *Thin Solid Films*, 193/194, 77 (1990); also P.A. Lindfors and W.M. Mularie, "Cathodic Arc Deposition Technology," *Surf. Coat. Technol.*, 29, 275 (1986); also A. Anders, "Approaches to Rid Cathodic Arc Plasmas of Macroparticles and Nanoparticles: A Review," *Surf. Coat. Technol.*, 120/121, 319 (1999)

239. André Anders, private communication

240. D.M. Sanders, "Ion Beam Self-Sputtering Using a Cathodic Arc Ion Source," *J. Vac. Sci. Technol.*, A6(3) 1929 (1987)

241. I.I. Aksenov, Y.P. Antifiv, V.G. Bren, V.A. Padalka, A.I. Popov, and A.I. Khoroshikh, "Effects of Electron Magnetization in Vacuum-Arc Plasma on the Kinetics of the Synthesis of Nitrogen-Containing Coatings," *Sov. Phys. Tech. Phys.*, 26(2) 184 (1981); also D.M. Sanders and E.A. Pyle, "Magnetic Enhancement of Cathodic Arc Deposition," *J. Vac. Sci. Technol.*, A5(4) 2728 (1987)

242. Richard M. Poorman and Jack L. Weeks, "Vacuum Vapor Deposition," U.S. Patent #5,380,415 (Jan. 10, 1995); also Jack L. Weeks and Douglas M. Todd, "Enhanced Vacuum Arc Vapor Deposition Electrode," U.S. Patent #5,891,312 (April 6, 1999); also loc. cit. [201]

243. I.G. Kesaev and V.V. Pashkova, "The Electromagnetic Anchoring of the Cathode Spot," *Sov. Phys. Tech. Phys.*, 3, 254 (1959) [English translation of *Zh. Tekh. Fiz.*, 29, 287 (1959)]

244. S. Ramalingham, U.S. Patent #4,673,477 (1987)

245. I.G. Brown, X. Godechot, and K.M. Yu, *Appl. Phys. Lett.*, 58, 1392 (1991); also *Handbook of Plasma Immersion Ion Implantation and Deposition*, edited by A. Anders, John Wiley (2000)

246. W.M. Conn, *Phys. Rev.*, 79, 213 (1950); also D.M. Mattox, A.W. Mullendore, and F.N. Rebarchik, "Film Deposition by Exploding Wires," *J. Vac. Sci. Technol.*, 4, 123 (1967)

247. *Chemical Vapor Deposition*, edited by Jong-Hee Park and T.S. Sudarshan, Surface Engineering Series, Vol. 2, ASM International (2001)

248. *Plasma Deposition, Treatment and Etching of Polymers*, edited by Riccardo d'Agostino, Academic Press (1990)

249. *Plasma Deposited Thin Films*, edited by J. Mort and F. Jansen, CRC Press (1986)

250. W.E. Sawyer and A. Man, U.S. Patent #229,335 (June 29, 1880)

251. L. Mond, U.S. Patent #455,230 (June 30, 1891)

252. J.W. Aylesworth, U.S. Patent # 553,296 (Jan. 21,

1896); also A. deLodyguine, U.S. Patent #575,002 (Jan,12, 1897); also A.E. van Arkel and J.H. de Boer, *Z. Anorg. Allg. Chem.,* 148, 345 (1925)

253. J.J. Lander and L.H. Germer, *Trans. AIME,* 175, 648 (1948); also H.E. Hintermann, A.J. Perry and E. Horvath, "Chemical Vapour Deposition Applied in Tribology," *Wear,* 47, 407 (1978)

254. J.C. Theurer, *J. Electrochem. Soc.,* 108, 649 (1961); also W.E. Spear and P.G. LeComber, *Solid State Comm.,* 17, 1193 (1975)

255. R.S. Berg and D.M. Mattox, "Deposition of Metal Films by Laser-controlled CVD," *Proceedings of the 4th International Conference on Chemical Vapor Deposition,* edited by G.F. Wakefield and J.M. Blocher, Jr., p. 196, Electrochemical Society (1973); also Jerome Lemelson, "Chemical Reaction Apparatus and Method," U.S. Patent 4,702,808 (Oct. 27, 1987)

256. J. Sandor, *Electrochem. Soc. Ext. Abs.,* No. 96, 228 (1962); also W. Kern and G. L. Schnable, "Low-Pressure Chemical Vapor Deposition for Very Large-Scale Integration Processing—A Review," *IEEE Transactions on Electron Devices,* ED-26(4) 647 (1979); also R.S. Rosler, "Low Pressure CVD Production Processes for Poly, Nitride and Oxide," *Solid State Technol.,* 20(4) 63 (1977); also P. Singer, "Techniques of Low Pressure CVD," *Semicond. Intl.,* p. 72 (May 1984)

257. W. Kern and R.S. Rosler, "Advances in Deposition Processes for Passivation," *J. Vac. Sci. Technol.,* 14, 1082 (1977)

258. W.A. Bryant, *J. Cryst. Growth,* 35, 257 (1976); also T. Suntola, *Thin Solid Films,* 216 84 (1992)

259. H. Jeon, J.-W. Lee, Y.-D. Kim, D.-S. Kim and K.-S. Yi, "Study on the Characteristics of TiN Thin Film Deposited by the Atomic Layer Chemical Vapor Deposition Method," *J. Vac. Sci. Technol.,* A18(4) 1595 (2000)

260. "Plasma Deposition of Inorganic Thin Films," J.R. Hollahan and R.S. Rosler, Ch. 4-1 in *Thin Film Processes,* edited by J.L. Vossen and W. Kern, Academic Press (1978)

261. W. von Bolton, *Z. Electrochem.,* 17, 971 (1911); also O. Ruff, *Z. Anorg. Allg. Chem.,* 99, 73 (1917); also G. Tammann, *Z. Anorg. Allg. Chem.,* 155, 145 (1921)

262. R.L. Stewart, *Phys. Rev.,* 45, 488 (1934); also H. König and G. Helwig, *Z. Phys.,* 129, 491 (1951); also L. Holland and S.M. Ojah, *Thin Solid Films,* 38, L17 (1976)

263. H. Schmellenmeier, *Exp. Tech. Phys.,* 1, 49 (1953); also *Z. Phys. Chem.,* 205, 349 (1955-1956); also B.V. Spitsyn, L.L. Bouilov, and B.V. Derjaguin, *J. Crys. Growth,* 52, 219 (1981)

264. "Technology of Vapor Phase Growth of Diamond Films," Richard J. Koba, Ch. 4 in *Diamond Films and Coatings,* edited by Robert F. Davis, William Andrew Publishing/Noyes Publications (1993); also J.C. Angus in *Diamond and Diamond-Like Films,* edited by J. Dismukes et al, *Electrochemical Society Proceedings,* Vol. PV 89-12, Electrochemical Society (1989); C.V. Deshpansey and R.F. Bunshah, "Diamond and Diamond-Like Films: Deposition Processes and Properties," *J. Vac. Sci. Technol.,* A7(3) 2294 (1989); also A. Altshuler, *Diamond Technology in the USSR: From Synthesis to Application,* Delphic Associates, (1990)

265. W.G. Eversole, U.S. Patent #3,030,188 (April 17, 1962); also D.J. Poferi, N.C. Gardner, and J.C. Angus, *J. Appl. Phys.,* 44(4) 1428 (1973)

266. A.R. Badzian and R.C. DeVries, *Mater. Res. Bull.,* 23, 385 (1988); also translation of Russian work (1970s) in *Surf. Coat. Technol.,* 38(1-2), (October 1989)

267. H.F. Sterling and R.C.G. Swann, *Solid State Technol.,* 6, 653 (1965); also M.J. Rand "Plasma-Promoted Deposition of Thin Inorganic Films," *J. Vac. Sci. Technol.,* 16(2) 420 (1979)

267a. S. Veprek and M. Heintz, *Plas. Chem. Plas. Proc.,* 10(1) 3 (1990)

267b. A.O. Klimovskii, A.V. Bavin, V.S. Tkalich, and A.A. Lisachenko, *React. Kinet. Catalysis Lett.* (from the Russian) 23(1-2)95 (1983)

268. "The Formation of Particles in Thin-Film Processing Plasmas," Christoph Steinbrüchel, p. 289 in *Plasma Sources For Thin Film Deposition and Etching: Physics of Thin Films,* Vol. 18, edited by Maurice H. Francombe and John L. Vossen, Academic Press (1994)

269. A.R. Reinberg, "Radial Flow Reactor," U.S. patent #3,757,733 (Sept. 11, 1973); also A.R. Reinberg, Abs. #6, *Extended Abstracts,* the Electrochemical Society Meeting, San Francisco, CA, May 12–17 (1974); also R.S. Rosler, W.C. Benzing, and J. Baldo, "A Production Reactor for Low-Temperature Plasma-Enhanced Silicon Nitride Deposition," *Solid State Technol.,* 19, 45 (June 1976); also A.R. Reinberg, *Ann. Rev. Mater. Sci.,* 9, 341 (1979); also M.J. Rand, *J. Vac. Sci. Technol.,* 16, 420 (1979)

270. L. Martinu and D. Poitras, "Plasma Deposition of Optical Films and Coatings: A Review," *J. Vac. Sci. Technol.,* A18(6) 2619 (2000)

271. D.M. Mattox, "Apparatus for Coating a Cathodically Biased Substrate From a Plasma of Ionized Coating Material," U.S. Patent #3,329,601 (July 4, 1967), assigned to the U.S. Atomic Energy Commission (USAEC)

272. "High Strength Ceramic-Metal Seals Metallized at Room Temperature," R. Culbertson and D.M. Mattox, p. 101 in *Proceedings of the 8th Conference on Tube Technology*, IEEE Conf. Record (1966)

273. Robert D. Culbertson, Russell C. McRae, and Harold P. Meyn, U.S. Patent #3,604,970 (1971)

274. W.E. Spear and P.G. LeComber, *Solid State Commun.*, 17, 1198 (1975)

275. D.E. Carlson and C.R. Wronski, *Appl. Phys. Lett.*, 28, 671 (1976)

276. P.G. LeComber, P.G. Spear, and A. Ghaith, *Electronic Lett.*, 15, 179 (1979); also W.E. Spear, *Adv. Phys.*, 26, 811 (1977); also W.E. Spear, p. 1 in *Amorphous and Liquid Semiconductors*, edited by J. Stude and W. Brenig, Taylor & Francis (1974)

277. T. Môri and Y. Namba, "Hard Diamondlike Carbon Films Deposited by Ionized Deposition of Methane Gas," *J. Vac. Sci. Technol.*, A1, 23 (1983); also S.R. Kasi, H. Kang, and J.W. Rabalais, "Chemically Bonded Diamond-like Films From Ion Beam Deposition," *J. Vac. Sci. Technol.*, A6(3) 1788 (1988); also S. Aisenberg and R.W. Chabot, "The Role of Ion-Assisted Deposition in the Formation of Diamond-like-Carbon Films," *J. Vac. Sci. Technol.*, A8(3) 2150 (1990)

278. K.S. Mogensen, C. Mathiasen, S.S. Eskildsen, H. Störi, and J. Bothiger, *Surf. Coat. Technol.*, 102, 35 (1998); also T.A. Beer, J. Laimer, and H. Störi, "Study of the Ignition Behavior of a Pulsed DC Discharge Used for Plasma-Assisted Chemical Vapor Deposition," *J. Vac. Sci. Technol.*, A18(2) 423 (2000)

279. "Chemical Beam Epitaxy," T.H. Chiu, Sec. A2.3.1 in *Handbook of Thin Film Process Technology*, edited by David Glocker and S. Ismat Shah, Institute of Physics Publishing (IOP) (1995)

280. W. Ensinger, *Rev. Sci. Instrum.*, 63, 5217 (1972); also *Large Ion Beams: Fundamentals of Generation and Propagation*, A. Theodore Forrester, p. 191, Wiley Interscience (1988)

281. S. Shanfield and R. Wolfson, "Ion Beam Synthesis of Cubic Boron Nitride," *J. Vac. Sci. Technol.*, A1(2) 323 (1983); also Hisao Katto, Kazunari Kobayashi, Yasushi Koga, and Machiko Koyama, "Method of Manufacturing a Metallic Oxide Film on a Substrate," U.S. Patent #3,668,095 (1972)

282. J. Madocks, "High-rate PECVD on Web Accomplished with New Magnetically Enhanced Source," p. 451 in *Proceedings of the 45th Annual Technical Conference*, Society of Vacuum Coaters (2002)

283. P. DeWilde, *Ber. Dtsch. Chem. Ges.*, 7, 4658 (1874)

284. A. Thenard, *C. R. Hebd. Seances Acad. Sci.*, 78, 219 (1874); also H. Hoenig and G. Helwig, *Z. Phys.*, 129, 491 (1951); also "Glow Discharge Polymerization," H. Yamada, Ch. IV-2, p. 361 in *Thin Film Processes*, edited by John L Vossen and Werner Kerm, Academic Press (1978); also C.J. Arguette, U.S. Patent #3,061,458 (1962); also *Plasma Polymerization*, H. Yasuda, Academic Press (1985)

285. H.P. Schreiber, M.R. Wertheimer, and A.M. Wróbel, "Corrosion Protection by Plasma-Polymerized Coatings," *Thin Solid Films*, 12, 487 (1980); also A.M. Wróbel and M.R. Wertheimer, "Plasma-Polymerized Organosilicones and Organometallics," p. 163 in *Plasma Deposition, Treatment, and Etching of Polymers*, edited by Riccardo d'Agostino, Academic Press (1990)

286. U.S. Patent # 6,156,566 (2001); also U.S. Patent #6,279,505 (2001)

287. "Plasma Polymer-Metal Composite Films," H. Biedermann and L. Martinu, p. 269 in *Plasma Deposition, Treatment, and Etching of Polymers*, edited by Riccardo d'Agostino, Academic Press (1990); also M. Neumann and M. Krug, "Plasma Assisted High-Rate Evaporation of Organically Modified SiO_2 Onto Plastics" (translated and edited by S.W. Schulz), *Vac. Technol. Coat.*, 3(3) 24 (2002); also German Patent Application #DE 195 48 160 C1

288. M. Dhayal, M.R. Alexander, and J.W. Bradley, "Investigating the Effect of Ion Flux and Electron Temperature for Plasma Polymerization of Acrylic Acid," to be presented at the 46th Annual Technical Conference of the Society of Vacuum Coaters (2003)

289. "Ion Plating and Ion Beam Assisted Deposition," Chapter 8 in *Handbook of Physical Vapor Deposition (PVD) Processing*, Donald M. Mattox, William Andrew Publishers/Noyes Publications (1998)

290. D.M. Mattox, "Ion Plating: Past, Present and Future," *Surf. Coat. Technol.*, 133, 517 (2000)

291. *Ion Plating Technology: Developments and Applications*, N.A.G. Ahmed, John Wiley (1987)

292. "Ion Plating," Donald M. Mattox, Ch. 6 in *Deposition Technologies for Films and Coatings*, second edition, edited by Rointan Bunshah et al, William Andrew Publishing/Noyes Publications (1994); there are extensive early references to ion plating in the "Ion Plating" chapter of the first edition of this book (1982)

293. Don Mattox, "Oral History Interview," Society of Vacuum Coaters (2002); also in 1995 Donald M. Mattox was awarded The Albert Nerken Award by the American Vacuum Society (now AVS International) *"for his invention of the ion-plating process and its continued development."*

294. *Ion Implantation and Plasma Assisted Processes*,

edited by Robert F. Hochman, Hillary Solnick-Legg, and Keith Legg, ASM International (1988)

295. A. Anders, "Ion Plating and Beyond: Pushing the Limits of Energetic Deposition," p. 360 in *Proceedings of the 45th Annual Technical Conference*, Society of Vacuum Coaters (2002)

296. D.M. Mattox, "Film Deposition Using Accelerated Ions," Sandia Corp. Development Report SC-DR-281-63 (1963); also "Film Deposition Using Accelerated Ions," *Electrochem. Technol.*, 2, 295 (1964); also "Plating with a Permanence," *Time* Magazine, 85(3) 58 (Jan. 15, 1965)

297. H.E. Farnsworth, R.E. Schlier, T.H. George, and R.M. Burger, *J. Appl. Phys.*, 26, 252 (1955); also *J. Appl. Phys.*, 29, 1150 (1958); also G.S. Anderson and Roger M. Moseson, "Method and Apparatus for Cleaning by Ionic Bombardment," U.S. Patent #3,233,137 (filed Aug. 28, 1961) (Feb.1, 1966); also "The Historical Development of Controlled Ion-assisted and Plasma-assisted PVD Processes," D.M. Mattox, p. 109 in *Proceedings of the 40th Annual Technical Conference*, Society of Vacuum Coaters (1997)

298. J.P. Coad and R.A. Dugdale, "Sputter Ion Plating," p.186 in *Proceedings of International Conference: Ion Plating & Allied Techniques (IPAT 79)* (London,) Published by CEP Consultants Ltd, Edinburgh (July 1979)

299. T. Ohmi and T. Shibata, "Advanced Scientific Semiconductor Processing Based on High-Precision Controlled Low-Energy Ion Bombardment," *Thin Solid Films*, 241, 159 (1993); also loc cit [62]

300. D.M. Mattox, Discussion Section of the *J. Electrochem. Soc.*, 115(12) 1255 (1968)

301. *Ionized Physical Vapor Deposition*, edited by Jeffery A. Hopwood, Academic Press (2000)

302. J.S. Colligon, "Energetics Condensation: Processes, Properties and Products," *J. Vac Sci Technol.*, A13, 1649 (1995); also O.R. Monterio, "Thin Film Synthesis by Energetic Condensation," *Annual Rev. Mat. Sci.*, 31, 111 (2001)

303. loc. cit. [251, figure 4]; also R.T. Johnson and D.M. Darsey, "Resistive Properties of Indium and Indium-Gallium Contacts to CdS," *Solid State Electronics*, 11, 1015 (1968)

304. D.M. Mattox and G.J. Kominiak, "Structure Modification by Ion Bombardment During Deposition," *J. Vac. Sci. Technol.*, 9, 528 (1972); also S. Aisenberg and R.W. Chabot, *J. Vac. Sci. Technol.*, 10, 104 (1973); also R.D. Bland, G.J. Kominiak, and D.M. Mattox, "Effect of Ion Bombardment During Deposition on Thick Metal and Ceramic Deposits," *J. Vac. Sci. Technol.*, 11, 671 (1974)

305. A.G. Blackman, *Metall. Trans.*, 2, 699 (1971); also D.W. Hoffman and J.A. Thornton, "Internal Stresses in Sputtered Chromium," *Thin Solid Films*, 40, 355 (1977)

306. G.A. Lincoln, M.W. Geis, S. Pang, and N. Efremow, *J. Vac. Sci. Technol.*, 13(1), 1043 (1983); also J.M.E. Harper, J.J. Cuomo, and H.T.G. Hentzell, *Appl. Phys. Lett.*, 43, 547 (1984): also Y. Itoh, S. Hibi, T. Hioki, and J. Kawamoto, *J. Mat. Res.*, 6, 871 (1991)

307. K.D. Kennedy, G.R. Schevermann, and H.R. Smith, Jr., "Gas Scattering and Ion Plating Deposition Methods," *R&D Mag.*, 22(11) 40 (1971)

308. "Improvements in and Relating to the Coating of Articles by Means of Thermally Vaporized Material," B. Berghaus, U.K. Patent #510,993 (1938); also B. Berghaus and W. Burkhardt, U.S. Patent #2,305,758 (December 1942)

309. R. Frerichs, "Superconductive Films by Protected Sputtering of Tantalum or Niobium," *J. Appl. Phys.*, 33, 1898, 1962

310. L.I. Maissel and P.M. Schaible, "Thin Films Formed by Bias Sputtering," *J. Appl. Phys*, 36, 237 (1965)

311. "Bias Sputtering of Molybdenum Films," R. Glang, R.A. Holmwood, and P.C. Furois, p. 643 in *Transactions of the 3rd International Vacuum Congress* (1965)

312. J.L. Vossen and J.J. O'Neill, Jr., "DC Sputtering with RF-Induced Substrate Bias," *RCA Review*, 29, 566 (1968); also John L. Vossen, Jr., "Method of Metallizing Semiconductor Devices," U.S. Patent #3,640,811 (Feb. 8, 1972)

313. L.I. Maissel, R.E. Jones, and C.L. Standley, *IBM Res. Dev.*, 14, 176 (1970); also R.D. Bland, G.J. Kominiak, and D.M. Mattox, "Effect of Ion Bombardment During Deposition on Thick Metal and Ceramic Deposits," *J. Vac. Sci. Technol.*, 11, 671 (1974)

314. E. Krikorian, "Preparation and Properties of Compound Materials by Reactive and Direct Sputtering," p. 38 in *Transactions 6th Conference and School on Applications of Sputtering* (1969) Sponsored by MRC; also J. Pompei, "The Mechanisms of DC, Modulated and RF Reactive Sputtering," p. 165 in *Transactions of the 3rd Symposium on Deposition of Thin Films by Sputtering* (1969), co-sponsored by Bendix and Univ. of Rochester; also R.H.A. Beal and R.F. Bunshah, p. 238 in *Proceedings of the 4th International Conference on Vacuum Metallurgy* (June 1973); also R.F. Bunshah and A.C. Raghuram, *J. Vac. Sci. Technol.*, 9, 1385 (1972); also M. Kobayashi and Y. Doi, *Thin Solid Films*, 54, 57 (1978)

315. D.M. Mattox and R.D. Bland, "Aluminum Coating

of Uranium Reactor Parts for Corrosion Protection," *J. Nucl. Materials*, 21, 349 (1967); also R.D. Bland, J.E. McDonald, and D.M. Mattox, "Ion Plated Coatings for the Corrosion Protection of Uranium," Sandia Corp. Report No. SC-R-65-519 (1965)

316. T. Spalvins, J.S. Przbyszewski, and D.H. Buckley, "Deposition of Thin Films by Ion Plating on Surfaces Having Various Configurations," NASA Lewis TN-D3707 (1966); also T. Spalvins, *J. Am. Soc. Lub. Eng.*, 33, 40 (1971); also B.C. Stupp, "Synergistic Effects of Metals Co-Sputtered With MoS_2," *Thin Solid Films*, 84, 257 (1981)

317. "Specialized Applications of Vapor-Deposited Films," L.E. McCrary, J.F. Carpenter, and A.A. Klein, *Proceedings of the 1968 Metallurgy Conference*, American Vacuum Society (1968); also D.M. Mattox and F.N. Rebarchik, "Sputter Cleaning and Plating Small Parts," *Electrochem. Technol.*, 6, 374 (1968); also K.E. Steube and L.E. McCrary, "Thick Ion-Vapor-Deposited Aluminum Coatings for Irregularly Shaped Aircraft and Spacecraft Parts," *J. Vac. Sci. Technol.*, 11, 362 (1974); also James F. Carpenter, Leon E. McCrary, Kenneth E. Steube, Albert A. Klein, Jr., and George H. Kesler, "Glow Discharge Coating Apparatus," U.S. Patent #3,750,623 (Aug. 7, 1973); also "Aluminum Plating by Ion Vapor Deposition," NAVMIRO (Naval Material Industrial Resources Office) Bulletin #72 (June 1977)

318. Military Specification MIL-C-83488

319. R.T. Bell and J.C. Thompson, "Applications of Ion Plating in Metal Fabrication," Oak Ridge, Y-12 Plant, Y-DA-5011 (1973); also J.W. Dini, "Ion Plating Can Improve Coating Adhesion," *Metal Finish.*, 80(9) 15 (1993)

319a. J.K.G. Panitz and D.J. Sharp, "The Effect of Different Alloy Surface Compositions on Barrier Anodic Film Formation," *J. Electrochem. Soc.*, 13(10) 2227 (1984)

320. D.M. Mattox and F.N. Rebarchik, "Sputter Cleaning and Plating Small Parts," *Electrochem. Technol.*, 6, 374, 1968

321. "Ion Plating Technology," D.M. Mattox, Ch. 6 in *Deposition Technologies for Films and Coatings—Development and Applications*, edited by R.F. Bunshah et al, Noyes Publications (1982); also J.N. Matossian, R.W. Schumacher, and D.M. Pepper, "Surface Potential Control in Plasma Processing of Materials," U.S. Patent #5,374,456 (Dec. 20, 1994)

322. S. Schiller, U. Heisig, and K. Goedicke, "Alternating Ion Plating—A Method of High-Rate Ion Vapor Deposition," *J. Vac. Sci. Technol.*, 12, 858 (1975)

323. P.M. Lefebvre, James W. Seeser, Richard Ian Seddon, Michael A. Scobey, and Barry W. Manley "Process for Depositing Optical Thin Films on Both Planar and Non-Planar Substrates" (filed July 7, 1994) U.S. Patent #5,798,027 (Aug. 25, 1998)

323a. N. Boling, B. Wood, and P. Morand, "A High Rate Reactive Sputtering Process for Batch, In-Line or Roll Coaters," p. 286 in *Proceedings of the 38th Annual Technical Conference*, Society of Vacuum Coaters (1995)

324. H.R. Barker and R.J. Hill, "The Deposition of Multicomponent Phases in Ion Plating," *J. Vac Sci. Technol.*, 9(6) 1395 (1972)

325. "Low Temperature Optical Coatings With High Packing Density Produced With Plasma Ion Assisted Deposition," S. Beisswenger, R. Gotzelmann, K. Matl, and A. Zoller, p. 21 in *Proceedings of the 37th Annual Technical Conference*, Society of Vacuum Coaters (1994); also S. Kadlec, J. Müsil, and W.-D. Müntz, "Sputtering Systems With Magnetically Enhanced Ionization for Ion Plating of TiN Films," *J. Vac. Sci. Technol.*, A8(3) 1318 (1990)

326. D.G. Teer, B.L. Delcea, and A.J. Kirkham, *J. Adhesion*, 8, 171 (1976)

327. B.A. Probyn, *Brit. J. Appl. Phys. (J. Phys. D.)* 1, 457 (1968); also Arthur Laudel, "Ion Beam Deposition System," U.S. Patent #3,583,361 (1971); also Gerald W. White, "New Applications of Ion Plating," p. 43, *R&D Mag.* (July 1973); also Y. Murayama, "Thin Film Formation of In_2O_3, TiN, and TaN by RF Reactive Ion Plating," *J. Vac. Sci. Technol.*, 12, 818 (1975)

328. A. Rockett, S.A. Barnett, and J.E. Greene, *J. Vac. Sci. Technol.*, B2, 306 (1984); also "Thin Film Deposition and Dopant Incorporation by Energetic Particle Sources," Samuel Strite and Hadis Morkoç, Section A2.4 in *Handbook of Thin Film Process Technology*, edited by David A. Glocker and S. Ismat Shah, Institute of Physics Publishing (IOP) (1995)

329. P.J. Martin, R.P. Netterfield, D.R. McKenzie, I.S. Falconer, C.G. Pacey, P. Tomas, and W.G. Saintly, "Characterization of a Titanium Arc and the Structure of Deposited Ti and TiN Films," *J. Vac. Sci. Technol.*, A(5), 22 (1987)

329a. W. Olbrich, J. Fessmann, G. Kampschulte and J. Ebberink, "Improved Control of TiN Coating Properties Using Cathodic Arc Evaporation With a Pulsed bias," *Surf. Coat. Technol.*, 49,258 (1991); also J. Fessmann, W. Olbrich, G. Kampschulte and J. Ebberink, "Cathodic Arc Deposition of TiN and Zr(C,N) at Low Substrate Temperatures Using a Pulsed Bias Voltage," *Mater. Sci. Eng.*, A140, 830 (1991)

330. D.M. Mattox and G.J. Kominiak, "Incorporation of

Helium in Deposited Gold Films," *J. Vac Sci. Technol.*, 8, 194 (1971); also J.J. Cuomo and R.J. Gambino, "Incorporation of Rare Gases in Sputtered Amorphous Metal Films," *J. Vac. Sci. Technol.*, 14, 152 (1977); also D. Chleck, R. Maehl, O. Cucchiara and E. Carnevale, *Int. J. Appl. Radiation Isotopes*, 14, 581 (1963); also D. Chleck and R. Maehl, *Int. J. Radiation Isotopes*, 14, 593 (1963); also D. Chleck and O. Cucchiara, *Int. J. Radiation Isotopes*, 14, 599 (1963); also [44]

331. P.J. Martin, R.P. Netterfield, W.G. Saintly, G.J. Clark, W.A. Lanford, and S.H. Sie, *Appl. Phys. Lett.*, 43, 711 (1983); also P.J. Martin, R.P. Netterfield, and W.G. Saintly, *J. Appl. Phys.*, 55, 235 (1984); also "Ion-Beam-Assisted Deposition," Graham K. Hubler and James K. Hirvonen, p. 593 in *Surface Engineering*, Vol. 5, ASM Handbook (1994)

332. L. Pranevicius, "Structure and Properties of Deposits Grown by Ion-Beam-Activated Vacuum Deposition Techniques," *Thin Solid Films*, 63, 77 (1979); also C. Weissmantel, G. Reisse, H.-J. Erler, F. Henny, K. Bewilogua, U. Ebersbach and C. Schurer, "Preparation of Hard Coatings by Ion Beam Methods," *Thin Solid Films*, 63, 315 (1979); also J.M.E. Harper, J.J. Cuomo, R.J. Gambino, and H.E. Kaufmann, Ch. 4 in *Ion Bombardment Modification of Surfaces: Fundamentals and Applications*, edited by O. Auciello and R. Kelly, Elsevier (1984); also Hans K. Pulker and Helmut Daxinger, "Method of Producing Gold-Color Coatings," U.S. Patent #4,254,159 (Mar. 3, 1981); also J.K. Hirvonen, "Ion Beam Assisted Thin Film Deposition," *Mat. Sci. Reports*, 6, 215 (1991)

333. S. Aisenberg and R. Chabot, "Study of the Deposition of Single-Crystal Silicon, Silicon Dioxide and Silicon Nitride on Cold-Substrate Silicon," Final Report prepared for NASA Electronic Research Center under Contract No. NAS 12-541 (October 1969); also *J. Appl. Phys.*, 42, 2953 (1971); also S. Aisenberg, U.S. Patent #3,904,505 (Sept. 9, 1975); also E.G. Spenser, P.H. Schmidt, D.C. Joy, and F.J. Sansalone, *Appl. Phys. Lett.*, 29, 118 (1976); also I.I. Aksenov, S.I. Vakula, V.G. Padalka, V.E. Strel'nitskii, and V.M. Khoroshikh, "High-Efficiency Source of Pure Carbon Plasma," *Sov. Phys. Tech. Phys.*, 25(9) 1164 (1980)

334. S. Aisenberg and R.W. Chabot, "Physics of Ion Plating and Ion Beam Deposition," *J. Vac. Sci. Technol.*, 10(1) 104 (1973); also Sol Aisenberg, "Apparatus for Film Deposition," U.S. Patent #3,904,505 (Sept. 9, 1975); also Sol Aisenberg, "Film Deposition," U.S. Patent # 3,961,103 (June 1, 1976); also S. Aisenberg and F.M. Kimock, p. 67 in *Preparation and Characterization of Amorphous Carbon Films*, edited by J.J. Pouch and S.A. Altervitz, Trans Tech

Publications (1990); also "Preparation, Structure, and Properties of Hard Coatings on the Basis of i-C and i-BN," C. Weissmantel, Ch. 4, p. 153 in *Thin Films From Free Atoms and Particles*, edited by Kenneth J. Klabunde, Academic Press (1985)

335. P.D. Prewett and C. Mahoney, "Applications of Coatings Produced by Field Emission Deposition," *Vacuum,* 43(3-4) 385 (1984)

336. J.A. Thornton, "The Influence of Bias Sputter Parameters on Thick Copper Coatings Deposited Using a Hollow Cathode," *Thin Solid Films,* 40, 335 (1977)

337. W.H. Wayward, and A.R. Wolter, *J. Appl. Phys.,* 40, 2911 (1969)

338. "Recent Advances in Ion Plating," D.M. Mattox, p. 443 in *Proceedings of the Sixth International Vacuum Congress,* *(Japan J. Appl. Phys. Suppl. 2, Pt. 1)*, Kyoto (1974); also M. Koboyashi and Doi, *Thin Solid Films*, 54, 17 (1978); also R.F. Bunshah, *Thin Solid Films*, 80, 255 (1981)

339. J.R. Conrad and T. Castagna, "Plasma Source Ion Implantation for Surface Modification," *Bull. Am. Phys. Soc.,* 31, 1479 (1986); also J.R. Conrad, R.A. Dodd, S. Han, M. Madapura, J. Scheuer, K. Sridharan, and F.J. Worzala, "Ion Beam Assisted Coating and Surface Modification with Plasma Source Implantation," *J. Vac. Sci. Technol.*, A8(4) 3146 (1990); also I.G. Brown, X. Godechot, and K.M. Yu, *Appl. Phys. Lett.*, 58, 1392 (1991); also S.M. Malik, K. Sridharan, R.P. Featherstone, A. Chen, and J.R. Conrad, *J. Vac. Sci. Technol.*, B12, 843 (1994); also "Method and Apparatus for Plasma Source Ion Implantation," J. Conrad, U.S. Patent #4,764,394 (Jan. 20, 1987)

340. I.G. Brown, R.A. MacGill, and J. Galvin, "Apparatus for Coating a Surface With a Metal Utilizing a Plasma Source," U.S. Patent #5,013,578 (Dec. 11, 1989); also I.G. Brown, X. Godechot, and K.M. Yu, *Appl. Phys. Lett.,* 58, 1392 (1991); also A. Anders, "From Plasma Immersion Ion Implantation to Deposition: A Historical Perspective on Principles and Trends," *Surf. Coat. Technol.,* 156, 3 (2002); also W. Emsinger, J. Klein, P. Usedom, and B. Rauschenbach, "Characteristic Features of an Apparatus for Plasma Immersion Ion Implantation and Physical Vapor Deposition," *Surf. Coat. Technol.,* 93, 175 (1997); also "Survey of PIII&D Intellectual Property," Jesse Matossian, p. 683 in *Handbook of Plasma Immersion Ion Implantation and Deposition*, edited by A. Anders, John Wiley (2000)

341. P.X. Yan, S.Z. Yang, B. Li, and X.S. Chen, "High Power Density Pulsed Power Deposition of Titanium Carbonitride," *J. Vac. Sci. Technol.*, A14(1) 115 (1996)

342. *Handbook for Critical Cleaning*, edited by Barbara Kanesberg et al, CRC Press (2001)

343. "Preparation and Cleaning of Vacuum Surfaces," Donald M. Mattox, Chapter 4.9 in *Handbook of Vacuum Technology*, edited by Dorothy M. Hoffman, Bawa Singh, and John H. Thomas, III, Academic Press (1998)

344. D.M. Mattox, "Surface Effects on the Growth, Adhesion and Properties of Reactively Deposited Hard Coatings," *Surf. Coat. Technol.*, 81, 8 (1996)

345. R.R. Sowell, R.E. Cuthrell, R.D. Bland, and D.M. Mattox, "Surface Cleaning by Ultraviolet Radiation," *J. Vac. Sci. Technol.*, 11, 474 (1974)

346. R.S. Berg and G.J. Kominiak, "Surface Texturing by Sputter Etching." *J. Vac. Sci. Technol.*, 13(1) 403 (1976); also P. Sigmund, *J. Mater. Sci.*, 8, 1545 (1973)

347. S. Schiller, U. Heisig, and K. Steinfelder, *Thin Solid Films*, 33, 331 (1976)

348. "Thin Film Adhesion and Adhesive Failure—A Perspective," D.M. Mattox, p. 54, in *ASTM Proceedings of Conf. on Adhesion Measurement of Thin Films, Thick Films and Bulk Coatings*, ASTM-STP640 (Code 04-640000-25) (1978); also "Interface Formation During Thin Film Deposition," D.M. Mattox and J.E. McDonald, *J. Appl. Phys.*, 34, 2493, 1963; also "Adhesion and Deadhesion," Ch. 11 in *Handbook of Physical Vapor Deposition (PVD) Processing*, Donald M. Mattox, William Andrew Publishing/Noyes Publications (1998)

349. F.B. Koch, R.L. Meek, and D.V. McCaughan, "Implantation of Argon in SiO_2 Due to Backsputter Cleaning," *Extended Abst. 142nd National Meeting of the Electrochemical Society*, Abs. No. 250, Vol. 72-2 (1972); also H.R. Deppe, B. Hasler, and J. Hopfner, "Investigations on the Damage Caused by Ion Etching of SiO_2 Layers at Low Energy and High dose," Solid State Electronics, 20, 51 (1977)

350. J.J. Vossen, J.J. O'Neill, Jr., K.M. Finlayson, and J.E. Turner, *RCA Rev.*, 31, 293 (1970)

351. R. Kelley, "Bombardment-Induced Compositional Change With Alloys, Oxides, Oxysalts, and Halides," p. 91 in *Handbook of Plasma Processing Technology: Fundamental, Etching, Deposition and Surface Interactions*, Noyes Publications (1990) (Alloys)

352. G. Betz and G.K. Wehner, "Sputtering of Multicomponent Materials," Chapter 2 in *Sputtering by Particle Bombardment II: Sputtering of Alloys and Compounds, Electron and Neutron Sputtering, Surface Topography*, edited by R. Behrisch, Springer-Verlag (1983)

353. D.J. Sharp and J.A. Panitz, *Surf. Sci.*, 118, 429 (1982)

354. Reza Alani, Joseph Jones, and Peter Swann, "Chemically Assisted Ion Beam Etching (CAIBE)—A New Technique for TEM Specimen Preparation of Materials," p. 85 in *Specimen Preparation for Transmission Electron Microscopy of Materials II* edited by Ron Anderson, Materials Research Society Symposium Proceedings, Vol. 199 (1990)

355. J. Strong, *Rev. Sci. Instrum.*, 6, 97 (1935); also "The Cleaning of Glass," Ch. 4. in *The Properties of Glass Surfaces*, L. Holland, John Wiley (1964)

356. G.J. Kominiak and D.M. Mattox, *Thin Solid Films*, 40, 141 (1977)

357. H.F. Dylla, *J. Nucl. Mat.*, 93/94, 61 (1980); also H.F. Dylla, *J. Vac. Sci. Technol.*, A6(3) 1276 (1988)

358. J. Roth, "Chemical Sputtering," Ch. 3 in *Sputtering by Particle Bombardment II*, edited by R. Behrisch, Springer-Verlag (1983)

359. J.A. Kelber, "Plasma Treatment of Polymers for Improved Adhesion," p. 255 in *Adhesion in Solids*, edited by D.M. Mattox, J.E.E. Baglin, R.J. Gottschall, and C.D. Batich, MRS Symposium Proceedings, Vol. 119 (1988); also E.M. Liston, L. Martinu, and M.R. Wertheimer, "Plasma Surface Modification of Polymers for Improved Adhesion: A critical review," p. 3 in *Plasma Surface Modification of Polymers*, edited by M. Strobel, C. Lyons, and K.L. Mittal, VPS Publicatons (1994); also L.J. Gerenser, "Surface Chemistry for Treated Polymers," Section E3.1 in Suppl. 96/2 in *Handbook of Thin Film Process Technology*, edited by David B. Glocker and S. Ismat Shah, Institute of Physics Publishing (1995); also M.R. Wertheimer, L. Martinu, and E.M. Liston, "Plasma Sources for Polymer Surface Treatment," Section E3.0 in Suppl. 96/2 in *Handbook of Thin Film Process Technology*, edited by David B. Glocker and S. Ismat Shah, Institute of Physics Publishing (1995)

360. J. Gazecki, G.A. Sai-Halasz, R.G. Alliman, A. Kellock, G.L. Nyberg and J.S. Williams, "Improvement in the Adhesion of Thin Films to Semiconductors and Oxides Using Electron and Photon Irradiation," *Appl. Surf. Sci.*, 22/23, 1034 (1985)

361. K. Rossman, *J. Polymer Sci.*, 19, 141 (1956)

362. R.H. Hansen and H. Schornhorn, *J. Polymer Sci.: Polymer Lett.*, B4, 203 (1966); also H. Schornhorn, F.W. Ryan and R.H. Hansen, "Surface Treatment of Polypropylene for Adhesive Bonding," *J. Adhesion*, 2, 93 (1970); also R.R. Sowell, N.J. DeLollis, H.J. Gregory, and O. Montoya, "Effects of Activated Gas Plasma on Surface Characteristics and Bondability of RTV Silicone and Polyethylene,"

p. 77 in *Recent Advances in Adhesion*, edited by L.-H. Lee, Gordon and Breach (1973)

363. J.K. Hirvonen, *J. Vac. Sci. Technol.*, 15,1662 (1978); also L.E. Pope, F.G. Yost, D.M. Follstaedt, S.T. Picraux, and J.A. Knapp, "Friction and Wear Reduction of 440C Stainless Steel by Ion Implantation," p. 661 in *Ion Implantation and Ion Beam Processing of Materials*, edited by G.K.Hubler, O.W. Holland, C.R. Clayton, and C.W. White, Vol. 27, *MRS Symposium Proceedings*, North-Holland (1984); also I. Masaya, *Crit. Rev. Solid State/ Mater. Sci.*, 15(5) 473 (1989)

364. C.K. Jones and S.W. Martin, *Metal. Prog.*, 85, 94 (1964); also A. Leland, K.S. Fancey, and A. Matthews, *Surf. Eng.*, 7(3) 207 (1991)

365. W.L. Grube and J.G. Gay, "High-Rate Carburizing in a Glow Discharge Methane Plasma," *Metall. Trans.*, 9A, 1421 (1978)

366. "The Formation of Particles in Thin-Film Processing Plasmas," Christoph Steinbrüchel, p. 289 in *Physics of Thin Films, Vol. 18—Plasma Sources for Thin Film Deposition and Etching*, edited by Maurice H. Francombe and John L. Vossen, Academic Press (1994)

367. D. Chen and S. Hackwood, "Vacuum Particle Generation and the Nucleation Phenomena During Pumpdown," *J. Vac. Sci. Technol.*, A8(2) 933 (1990); also J.J. Wu, D.W. Cooper, and R.J. Miller, "Aerosol Model of Particle Generation During Pressure Reduction," *J. Vac. Sci. Technol.*, A8(3) 1961 (1990); also J.F. O'Hanlon, "Advances in Vacuum Contamination Control for Electronic Materials Processing,"*J. Vac. Sci. Technol.*, A5(4) 2067 (1987)

368. loc. sit. [2] p. 74; also [14]

368a. "The Low-Pressure Gas and Vacuum Processing Environment," Ch. 3, Table 3-11, in *Handbook of Physical Vapor Deposition (PVD) Processing*, Donald M. Mattox, William Andrew Publishing/ Noyes Publications (1998)

369. R.W. Phillips, "Optically Variable Films, Pigments, and Foils," SPIE Vol. 1323, *Optical Thin Films III: New Developments*, 98 (1990); also A. Argoitia and M. Witzman, "Pigments Exhibiting Diffractive Effects," p. 539, *Proceedings of the 45th Annual Technical Conference*, Society of Vacuum Coaters (2002)

370. *They All Laughed: From Light Bulbs to Lasers: The Fascinating Stories Behind the Great Inventions That Have Changed Our Lives*, Ira Flatow, HarperCollins (1992); also *The Last Lone Inventor: A Tale of Genius. Deceit & the Birth of Television*, (Philo T. Farnsworth vs. David Sarnoff), Evan I. Schwartz, HarperCollins (2002)

371. *Stress-Induced Phenomena in Metallization*, edited by P.S. Ho, C. Li, and P. Totta, AIP Conference Proceedings (1985)

372. J.W. Vossen, A.W. Stephens and G.L. Schnable, *Bibliography on Metallization Materials & Techniques for Silicon Devices*, AVS Publication (1974); also John Vossen, *Bibliography on Metallization Materials & Techniques for Silicon Devices—(II-IX)*, AVS Publication (1976-1983). These bibliographies were distributed free to members of the AVS Thin Film Division and were for sale to others.

373. *Procedures in Experimental Physics*, J. Strong, p. 168, Prentice-Hall (1938)

Acronyms Used in Vacuum Coating

Aa

a (α)	Amorphous (Example: a-Si)
A	Ampere
Å	Ångstrom
AAS	Atomic adsorption spectroscopy
ABS	Acrylonitrile-butadiene-styrene copolymer; Alky-benzene-sulfonate detergent
ACGIH	American Conference of Governmental Industrial Hygienists
a-C:H	Amorphous hydrogen-containing carbon (one form of DLC)
ACS	American Chemical Society
AEM	Analytical electron microscopy
AES	Auger electron spectroscopy
AESF	American Electroplaters and Surface Finishers
AF	Audio frequency
AFM	Atomic force microscopy; Abrasive flow machining
AIMCAL	Association of Industrial Metallizers, Coaters and Laminators, Inc.
AIP	American Institute of Physics
AMLCD	Active-matrix liquid crystal display
AMR	Anisotropic magnetoresistive
amu	Atomic mass unit
ANSI	American National Standards Institute
AO	Atomic oxygen
APC	Adaptive process control
APCVD	Atmospheric pressure chemical vapor deposition
APGD	Atmospheric pressure glow discharge
APIMS	Atmospheric pressure ionization mass spectrometry
APS	American Physical Society
AR	Antireflective
ARAS	Antireflective/antistatic
ARC	Antireflective coating
ARE	Activated reactive evaporation
ARIP	Activated reactive ion plating
ARO	After receipt of order

ASHRAE	American Society of Heating, Refrigerating and Air-Conditioning Engineers
ASIC	Application specific integrated circuit
ASM	ASM International (previously American Society for Metals; now ASM International)
ASME	American Society of Mechanical Engineers
ASNT	American Society for Nondestructive Testing
ASQC	American Society for Quality Control
ASTM	American Society for Testing and Materials
atm	Atmosphere (usually standard atmosphere)
at%	Atomic percent
AVEM	Association of Vacuum Equipment Manufacturers (more correctly known as AVEM International)
AVS	Society that used to be known as the American Vacuum Society

Bb

B	Magneic field (vector)
BAG	Bayard-Alpert gauge
BBAR	Broad band antireflection
bcc	Body centered cubic (crystallography)
BOPP	Biaxially oriented polypropylene
bp	Boiling point
BP	Bandpass (filter)
BPSG	Borophosphosilicate glass
BRDF	Bidirectional reflectance distribution function (light)
BSC	Black sooty crap

Cc

c	Velocity of light in a vacuum
C	Capacitance; Ceiling
CAD	Computer aided design
CAM	Computer aided manufacturing
CAPVD	Cathodic arc physical vapor deposition
CAS	Chemical abstract service

CASS	Copper accelerated acetic acid salt spray
cc	Cubic centimeter
CCAI	Chemical Coaters Association International
CCD	Charged-coupled devices
CCW	Counterclockwise
CD	Compact disc; Critical dimension; Cross direction
CDG	Capacitance diaphragm gauge
CDMS	Chlorodimethylsilane
CD-R	Compact disc-recordable
CEVC	Completely enclosed vapor cleaner
CF™	Conflat (vacuum flange)
CFC	Chlorofluorocarbon
CFC-111	Trichloroethane
CFC-113	Trichlorotrifluoroethane
cfm	Cubic feet per minute
cfs	Cubic feet per second
CGA	Compressed Gas Association
cgs	Centimeter–gram–second system of measurement
CIE	Commission International de l'Eclairage (International Comimssion on Illumination)
CLA	Center line average
CLEO	Conference on Laser and Electro-Optics
cm	Centimeter
CMM	Converting machinery/materials
CMOS	Complementary metal-oxide semiconductor
CMP	Chemical-mechanical polishing; Chemical-mechanical planarization
CN	Coordination number
CNDP	Cold neutron depth profile
COO (CoO)	Cost of ownership
CPP	Cast poproplylene
CrP	Chromium-rich oxide passivation
CRT	Cathode ray tube
CTE	Coefficient of thermal expansion
CTMS	Chlorotrimethylsilane
C-V	Capacitance-voltage
CVD	Chemical vapor deposition
CW	Clockwise

𝒟d

d	Day
D-CVD	Dielectric-CVD
DCS	Dichlorosilane
di-	2; Two
DI	De-ionized water
Diff	Diffusion pump
DIO	De-ionized and ozonated (water)
DIW	De-ionized water
DLC	Diamond-like-carbon
DMS	Dual magnetron sputtering
DMSO	Dimethyl sulfoxide
DOE	Department of Energy (U.S.); Design of experiments
DOP	Dioctyl phthalate
DOT	Department of Transportation
DOVID	Diffractive optically variable image device
DP	Diffusion pump
DRAM	Dynamic random access memory
DTIC	Defense Technical Information Center (USA)
DVD	Directed vapor deposition
DUV	Deep ultraviolet
DWDM	Dense wavelength division multiplexing
dwt	Pennyweight

ℰe

E	Emissivity; Electric field (vector); Exponential
EB (eb)	Electron beam
ECM	Electrochemical machining
ECR	Electron cyclotron resonance
ECS	Electrochemical Society
EDM	Electrodischarge machining
EDX	Energy dispersive X-ray
EDTA	Ethylene diamine tetraacetic acid
EELS	Electron energy loss spectroscopy
EHC	Electrolytic hard chrome
EL	Electroplated
ELD	Electroluminescent display (flat panel)
emf	Electromotive force
EMI	Electromagnetic interference
EN	Electroless nickel

EPA	Environmental Protection Agency
epi	Epitaxial
ERA	Evaporative rate analysis
ERD	Elastic recoil detection
ESCA	Electron spectroscopy for chemical analysis
ESD	Electrostatic discharge
eV	electron volt

Ff

F	Farad; Free machining (steel)
FC	Fault classification
fcc	Face centered cubic
FD	Fault detection
FDD	Floppy disc drive
FEC	Field emission cathode
FED	Field emission display; Field emission diode
FE-SEM	Field emission—scanning electron microscopy
FET	Field effect transistor
FLIR	Forward looking infrared (7.5 to 12 μm)
FPC	Fixed process control
FPD	Flat panel display
fpm	Feet per minute
FT-IR	Fourier transform infrared analysis

Gg

g	Unit of gravitational acceleration; gram
G	Giga (suffix for 10^9); unit of magnetic field strength (Gauss); Gallons
GANA	Glass Association of North America
GDMS	Glow discharge mass spectrometry
GDOES	Glow discharge optical emission spectroscopy
GFCI	Ground fault circuit interrupter
GLAD	Glancing angle deposition
GPM	Gallons per minute
gr	Grain

Hh

h	Planck's constant
H	Hour; Henry (unit of inductance)
HAP	Hazardous air pollutants

HAZ	Heat-affected zone; Hazardous (material)
HCD	Hollow cathode discharge
HCL	Hollow cathode lamp; Hydrochloric acid
hcp	Hexagonal close packed
HDD	Hard disk drive
HDP-CVD	High-density plasma CVD
HEED	High energy electron diffraction
HEPA	High efficiency particle air (see also ULPA)
HF	Hydrofluoric acid
HFCVD	Hot filament chemical vapor deposition
HFE	Hydrofluoroether
HMC	Hybrid micro circuit
HMCTSO	Hexamethylcyclotrisiloxane
HMDSO	Hexamethyldisiloxane
HRI	High refractive index
HVOF	High velocity oxygen fuel
HWOT	Half wave optical thickness
Hz	Hertz (cycles per second)

Ii

IAD	Ion assisted deposition
IARC	International Agency for Research on Cancer (establishes carcinogenicy of materials)
IBA	Ion beam analysis
IBAD	Ion beam assisted deposition
IBED	Ion beam enhanced deposition
IBEST™	Ion beam surface treatment
IC	Integrated circuit
ICB	Ionized cluster beam (deposition)
ICP	Inductively coupled plasma
ICP-MS	Inductively coupled plasma mass spectrometer
ID	Internal diameter
IDLH	Immediately dangerous to life or health
IDM	Integrated device manufacturing
IEEE	Institute of Electrical and Electronic Engineers
IES	Institute of Environmental Sciences
ILD	Interlayer dielectric
IMD	Intermetal dielectric
IMEMS	Integrated microelectromechanical systems
IPA	Isopropyl alcohol

IPC	Institute for Interconnecting and Packaging Electronic Circuits
iPVD	Ionized physical vapor deposition
I-PVD	Ion-assisted physical vapor deposition
ISHM	International Society for Hybrid Microelectronics
ISO	International Standards Organization
ISS	Ion scattered spectrometry
ITO	Indium-tin-oxide alloy (90 : 10)
I-V	Current–voltage
IVD	Ion vapor deposition
IWFA	International Window Film Association

Jj

J	Joule; Electric current (vector)
JVST	Journal of Vacuum Science and Technology

Kk

K	Dielectric constant; Karat (fineness of gold)
k	Kilo (suffix for 10^3); Boltzman's constant; Portion of the complex index of refraction given by n-ik or n(1-ik); Optical extinction coefficient (k=al/4p)
kcal	Kilocalorie
kGy	KiloGray
kWH	Kilo-watt-hour

Ll

l	Liter (not preferred)
L	Low (carbon steel); Liter (preferred)
LAD	Laser ablation deposition
LASER	Light-amplification by stimulated emission of radiation
LC $_{50}$	Median lethal dose
LCD	Liquid crystal display
LCM	Laser confocal microscope
LCVD	Laser chemical vapor deposition
LDPE	Low density polyethylene
LED	Light emitting diode
LEED	Low energy electron diffraction
LLDPE	Linear low density polyethylene
LM	Layer metallization

LOCOS	Local oxidation of silicon
LP-CVD	Low pressure chemical vapor deposition (see also SA-CVD)
LPPS	Low-pressure plasma spray
LIMA	Laser-induced mass analysis
LLS	Linear least squares (statistical analysis)
LN & LN2	Liquid nitrogen
LPCVD	Low-pressure chemical vapor deposition
LTEL	Long-term exposure limits
LTS	Long-throw sputtering
LWP	Long-wavelength pass filter

Mm

m	Milli (suffix for 10^{-3}); Molality
M	Mega (prefix for 10^6); Minute
MBE	Molecular beam epitaxy
mcg	Micrograms
MCrAlY	Metal-chromium-aluminum-Yitterium
MD	Movchan-Demchiskin; Machine direction
MDG	Molecular drag gauge
Me	Metal
Me-C:H	Metal-containing hydrocarbons
MEC	Methylene chloride
MEMS	Microelectromechanical systems (also called MST)
MePIIID	Metal plasma imersion ion implantation and deposition
MERIE	Magnetically enhanced reactive ion etcher
MF	Mid-frequency
MFC	Mass flow controller
MFM	Mass flow meter
MFSA	Metal Finishing Supplier's Association
min	Minute
mks, MKS	Meter–kilogram–second system of measurement
ML	Monolayer
MLAR	Multi-layer antireflection coating
MLS	Monolayers per second
MMIC	Monolithic microwave integrated circuits
MNS	Metal-nitride-silicon
MO	Magneto-optical
MOCVD	Metal-organic chemical vapor deposition
mono-	1; One
MOS	Metal-oxide semiconductor
MoS2M	Metal-containing MoS_2

MPI	Manufacturing process instruction
MR	Magnetoresistive
MRS	Materials Research Society
MSDS	Materials safety data sheet
MST	Microsystems technology (also called MEMS)
MT-CVD	Medium temperature chemical vapor deposition
MTR	Material test report
MVTR	Moisture vapor transmission rate

Nn

n	Index of refraction; Portion of the complex index of refraction given by n-ik
NACE	National Association of Corrosion Engineers
NAMF	National Association of Metal Finishers
NBS	National Bureau of Standards, which has been renamed NIST
nc	Nanocomposite
NC	Normally closed
NDE	Nondestructive evaluation
NDT	Nondestructive testing
NEG	Non-evaporable getter
NESHAP	National emission standards for hazardous air pollutants
NFPA	National Fire Protection Association
NIST	National Institute of Science and Technology (USA)
nm	Nanometer
NO	Normally open
NPB	N-propyl bromide
NREL	National renewable energy laboratory
NVR	Non-volatile residues

Oo

OD	Optical density; Outside diameter
ODP	Ozone depletion potential
OEM	Original equipment manufacturer
OES	Optical emission spectroscopy
OLED	Organic light emitting devices; Organic luminescent devices
OPP	Oriented polypropylene
OS	Ozone safe
OSEE	Optically stimulated electron emission

OSHA	Occupational Safety and Health Administration (USA)
OTR	Oxygen transmission rate
OVID	Optically variable image display
OXTR	Oxygen transmission rate
oza or oz(a)	Avoirdupois ounce
ozt or oz(t)	Troy ounce

Pp

p	Parallel (Example: p wave)
P	Suffix used to denote plasma deposited material (Example: P-TEOS)
Pa	Pascal
PA	Polyamide
PACVD	Plasma assisted chemical vapor deposition
PAPVD	Plasma assisted physical vapor deposition
PAVD	Plasma assisted vapor deposition
PC	Polycarbonate
PCE	Perchloroethylene
PD	Plasma doping
PDP	Plasma display panel
PECVD	Plasma enhanced chemical vapor deposition
PEEK	Polyethyletherketone
PEI	Polyetherimide
PEL	Permissible exposure limits
PEM	Plasma emission monitor
PEMS	Plasma enhanced magnetron sputtering
PERC	Perchloroethylene
PET	Polyethylene terephthalate (polyester)
penta-	5; Five
PF	Packing fraction
PFC	Perfluorocompounds
PFD	Process flow diagram
PFPE	Perfluorinatedpolyether
pH	Pouvoir hydrogene
PICVD	Plasma impulse CVD
PIII	Plasma immersion ion implantation
PLD	Pulsed laser deposition
PM	Preventive maintenance
PML	Polymer multilayer
PMS	Pulsed magnetron sputtering
PO	Purchase order
poly	Polycrystalline

POU	Point of use
PP	Polypropylene; Plasma polymerization
ppm	Parts per million
ppmbv	Parts per million by volume
PSG	Phosphosilicate glass
psi	Pounds per square inch
psia	Pounds per square inch—absolute
psig	Pounds per square inch—gauge
PVA	Polyvinyl alcohol
PVC	Polyvinyl chloride
PVD	Physical vapor deposition
PVDC	Polyvinylidene chloride
PWB	Printed wiring board
PZT	Lead zirconate titanate ($PbZrTiO_3$)

Qq

Q	Charge in coulombs
QA	Quality assurance
QC	Quality control
QCM	Quartz crystal monitor
QWOT	Quarter wavelength optical thickness

Rr

R	Resistance; Organic radical in chemical nomenclature
R_a	Roughness (average)
R_{max}	Roughness (maximum)
R_s	Sheet resistance; Spreading resistance
RAM	Random access memory
RBS	Rutherford backscattering spectrometry
RED	Reflection electron diffraction
rf	Radio frequency
RFI	Radio frequency interference
RFID	Radio frequency Identification
RFQ	Request for quotes
RGA	Residual gas analyzer
RH	Relative humidity
RHEED	Reflection high energy electron diffraction
RIBE	Reactive ion beam etching
RIE	Reactive ion etching
RMOS	Refractory metal-oxide semiconductor
rms	Root mean square
RO	Reverse osmosis
ROM	Read-only memory

ROW	Rest of world
RPE	Reactive plasma etching
rpm	Revolutions per minute
rps	Revolutions per second
RT	Room temperature
RTA	Rapid thermal annealing
RTCVD	Rapid thermal CVD
RTN	Rapid thermal nitridation
RTP	Rapid thermal processing
RTSPC	Real time statistical process control

Ss

s	Second; Perpendicular (as in s-wave)
SA-CVD	Sub-atmospheric CVD
SAD	Selected area diffraction
SAE	Society of Automotive Engineers
SAMPE	Society for the Advancement of Materials and Processing Engineering
SAW	Surface acoustic wave
SCBA	Self contained breathing apparatus
sccm	Standard cubic centimeters per minute
sccs	Standard cubic centimeters per second
scf	Standard cubic feet
scm	Standard cubic meters
SCM	Scanning capacitance microscope
SCSI	Small computer systems interface
SEAM	Scanning electron acoustic microscope
SEI	Secondary electron image
SEM	Scanning electron microscopy
SEMI	Semiconductor Equipment and Materials International
sg	Specifc gravity
SI	International System (system of units)
SIAM	Scanning interferometric apertureless microscope
SIMOX	Separation by implanted oxygen
SIMS	Secondary ion mass spectrometry
SION	Silicon oxynitride
SIP	Sputter ion plating
SIS	Semiconductor-insulator-semiconductor
SLAM	Scanning laser acoustic microscope
SLAR	Single layer antireflection
slm	Standard liters per minute
SMART	Self-monitoring analysis and reporting technology

SME	Society of Manufacturing Engineers
SMIF	Standard mechanical interface
SMT	Surface mount technology
SNMS	Secondary neutral mass spectrometry
SOD	Spin-on-dielectric
SOG	Spin-on-glass
SOI	Silicon-on-insulator
SPC	Statistical process control
SPE	Solid phase epitaxy
SPIE	International Society for Optical Engineering
SQUID	Superconducting quantum interference device
SRAM	Static random access memory
SRG	Spinning rotor gauge
SRM	Standard reference material
SS (SST)	Stainless steel
SSIS	Surface scanning inspection systems
SSMS	Spark source mass spectrometry
std	Standard
STEL	Short-term exposure limits
STEM	Scanning transmission electron microscopy
SThM	Scanning thermal microscopy
STI	Shallow trench isolation
STM	Scanning tunneling microscopy
STP	Standard temperature (0°C) and pressure (760 Torr)
SVC	Society of Vacuum Coaters
SWP	Short wavelength pass filter
SZM	Structure-zone-model

Tt

TA	Thermal analysis
TAB	Tape automated bonding
t:aC	Tetrahedral amorphous carbon
ta-C:H	Tetrahedral-bonded carbon (no hydrogen) (one form of DLC)
TA-MS	Thermal analysis with mass spectrometry
TC	Thermocouple; Thermocompression
TCA	1,1,1-trichloroethane (or methyl chloroform)
TCC	Transparent conductive coating
TCE	Trichloroethylene ($CHCl:CCl_2$); Thermal coefficient of expansion

TCLP	Toxicity characteristic leaching procedure
TCO	Transparent conductive oxide
TCP	Transformer-coupled plasma
TCR	Temperature coefficient of resistivity
TD	Transverse direction
TEM	Transmission electron microscopy
TEOS	Tetraethoxysilane
tetra-	4; Four
TFI	Thin-film inductive
TFT	Thin film transistor
TGA	Thermogravimetric analysis
TGA-MS	Thermogravimetric analysis with mass spectrometry
TIS	Total integrated scatter
TiW	(W:10wt%Ti) or (W:30at% Ti) (alloy)
TLV	Threshold limit values
TMDSO	Tetramethyldisiloxane
TMP	Turbomolecular pump
TMS	Tetramethyldisiloxane (TMDSO preferred)
tri-	3; Three
TSHT	Total solar heat transmittance
TWA	Time-weighted average
TWM	Thermal wave microscopy
TZM	Alloy of titanium, zirconium, and molybdenum

Uu

u	Unified atomic mass unit
UBM	Unbalanced magnetron
UCHF	Ultra-clean high flow
UF	Ultra-filtration
UHP	Ultra-high purity
UHV	Ultra-high vacuum
ULPA	Ultra-low permeation air
ULSI	Ultra-large scale integration
uPVC	Unplasticized polyvinyl chloride
UPW	Ultra-pure water
USPTO	US Patent and Trademark Office
UTS	Ultimate tensile strength
UV	Ultraviolet

Vv

v	Velocity
V	Volt; Voltage (as in CV measurements)
VAR	Vacuum arc remelting
VCR	Voltage coefficient of resistance
VEPA	Very-high efficiency particulate air (filter)
VHV	Very high vacuum
VIM	Vacuum induction melting
VLP-PECVD	Very-low-pressure plasma enhanced chemical vapor deposition
VLR	Visible light reflection
VLT	Visible light transmission
VOC	Volatile organic compounds
VOD	Vacuum oxygen decarburization
VPE	Vapor phase epitaxy
VUV	Vacuum ultraviolet

Ww

W	Watt
WDM	Wavelength division multiplexing
WDX	Wavelength dispersive X-ray
WORM	Write once read many
wt%	Weight percent
WVTR	Water vapor transmission rate

Xx

XES	X-ray energy spectroscopy
XPS	X-ray photoelectron spectroscopy
XRD	X-ray diffraction
XRF	X-ray fluorescence
XRM	X-ray microanalysis
XRT	X-ray topography
XUHV	Extra ultra-high vacuum

Yy

Y	Young's modulus

Zz

Z	Atomic number of an element
ZAO	Aluminum-doped zinc oxide
ZD	Zero defects

Miscellaneous Symbols

a	Optical adsorption coefficient (cm^{-1})
Ω	Ohm
μ	Micron
μm	Micrometer
n	Frequency
N	Normal
l	Wavelength
i	Prefix used to indicate that the film was formed using beam-type film-ion deposition. Example: i-C, i-BN.

Other Glossaries

ASTM Standard E 673-86a, "Definitions of Terms Relating to Surface Analysis"

"Scientific Unit Conversion," Francois Cardarelli. Springer-Verlag (1997)

Glossary of Terms for Vacuum Coating

A

Abnormal glow discharge (plasma)   The DC glow discharge where the cathode spot covers the whole cathode and an increase in the voltage increases the cathode current density. This is the type of glow discharge used in most plasma processing. See Normal glow discharge.

Abrasion test (characterization)   Testing film adhesion and abrasion resistance by rubbing, impacting, or sliding in contact with another surface or surfaces. Examples: **Tumble test; Tabor test; Eraser test.**

Abrasive (cleaning)   A material, such as a particle or a rough solid, that is capable of removing material from a surface when there is pressure and movement between the material and the surface.

Abrasive cleaning   The removal of surface material (gross cleaning), including contamination, by an abrasive action.

Abrasive compound   A material used to remove material from a surface by abrasion. Surface smoothness after abrasion is a secondary consideration. Examples: Silicon carbide; Emery; Silica; Alumina. See Polishing compound.

Abrasive flow machining (vacuum technology)   A means of smoothing a surface using a slurry of abrasive particles in a fluid that is passed over the surface. Also called **Slurry polishing**.

Abrasive transfer, contamination by (cleaning)   Transfer of material to a clean surface by contact or friction with a material to which it adheres such as a polymer on a high surface energy surface or chromium on a clean oxide surface.

Abrupt-type interface (film formation)   The interface that is formed between two materials (A and B) when there is no diffusion or chemical compound formation in the interfacial region. The transition of A to B in the length of a lattice parameter ($\approx$3Å). See Interface.

Absolute humidity   The amount of water vapor in the air as measured in grams per cubic centimeter.

Absorbate   The material being absorbed.

Absorption   Condition where the material on the surface (absorbate) diffuses into the bulk of the material (absorbent). See Adsorption.

Absorptivity (optics)   The absorption of radiation as it passes through a material. See Coefficient of extinction.

Abstraction (chemistry)   Removal of a species. Example: Abstraction of fluorine from a polymer surface.

Accelerated life test (adhesion)   A test conducted at a stress higher than that encountered in normal operation for the purpose of producing a measurable effect such as the loss of adhesion, in a shorter time than experienced at normal operating conditions. Examples: Elevated temperature; Concentrated chemical environment.

Acceleration due to gravity (g)   Acceleration equal to the standard acceleration due to gravity or 9.80665 meters per second per second.

Acceptor   An impurity (dopant) that decreases the number of free electrons in the material. See Donor.

Accuracy   The closeness of agreement between an observed value and an accepted reference value. See Precision.

Acetone (cleaning)   Solvent with the chemical formula CH_3COCH_3, also known as 2-propanone.

Acetylene (C_2H_2) (reactive deposition)   A hydrocarbon gas that is used as a chemical vapor precursor to provide carbon in reactive deposition processes.

Acid   Any chemical species capable of supplying a proton (hydrogen ion) to react with another chemical species. An acid yields hydrogen ions (H^+) by reaction with the solvent while a base forms hydroxyl ions, OH^-. See Lewis acid; Base; pH.

Acid pickling (cleaning)   Removal of the heavy oxide layer, such as a mill-scale, on a metal by acid etching.

Acidic surface (adhesion, film formation)   A surface capable of accepting an electron from an atom in contact with it. See Basic surface.

Acoustic   Relating to sound which is the transmission of a property, such as pressure, through a medium. Sound in the auditory range of the human ear ($\approx$30 Hz to 16 kHz) is called **Sonic**; above the auditory range (> 16 Hz) it is called **Ultrasonic**; and below the auditory range (< 30 Hz), **Infrasonic**.

Acoustic emission (adhesion)   The acoustic (sound) emission from a material being fractured or in some cases deformed.

Acoustic streaming (cleaning)   The currents in the fluid that are set up by the acoustic transmission through the fluid in ultrasonic cleaning. Capable of carrying particulates from the bottom of the tank into the cleaning area.

Actinometry (plasma technology)   Compares the emission interactions of the excited states of reference and subject species to obtain the relative concentrations of the ground states of the species.

Activated carbon   A form of carbon that has a very high surface area (> 1000 m²/g) due to the large number of fine pores in the material. Used to absorb vapors and organics in water purification. Can absorb gases when cooled (cryosorption). Can be regenerated (lose adsorbed gases) at room temperature.

Activated reactive evaporation (ARE) (PVD technology) Evaporation through a plasma of reactive gas in order to deposit a film of a compound material. The plasma activation increases the reaction probability and decreases the pressure of reactive gas needed to form the compound material.

Activation, plasma The process of making a species more chemically reactive by excitation, ionization, fragmentation, or forming new materials in a plasma.

Activation energy The energy barrier that isolates one chemical state from another as viewed from the reactant side.

Active film A film that will change properties (color, electron emission, optical transparency) under an externally applied stimulus (electric field, temperature, mechanical deformation). See Passive film.

Active gas A gas that will chemically react with an atom or molecule. Also called a **Reactive gas**. See Inert gas.

Active storage (cleaning) Storage in an environment that is continually being cleaned to remove potential contaminants. See Passive storage.

Adatom (film formation) The atom that has been deposited on the surface and that is still mobile (not condensed) on the surface.

Adatom mobility (film formation) The degree to which an adatom can move on the surface and condense at a nucleation site. The lower the mobility, the higher the nucleation density. See Nucleation density.

Addition agents (electroplating) Chemical agents added to the electroplating bath in order to influence some property of the deposited coating. Examples: **Brightening agents**; **Complexing agents**; **Leveling agents**; **Grain refiners**. Also called **Additives**.

Adhesion The physical bonding between the two surfaces of different materials. See Cohesion.

Adhesion, apparent The adhesion observed by applying an external force. If the internal stress is high the apparent adhesion may be low even though there is strong bonding at the interface because the internal stress adds to the applied external stress to cause failure. Also called **Practical adhesion**.

Adhesion failure Failure in the interfacial region (or near the interfacial region) by fracture or deformation. Also called **Deadhesion**.

Adhesion test A test to give an indication of the adhesion and to ensure product reproducibility and functionality. Often the adhesion test is used in a comparative manner to compare to previous findings.

Adhesion test, bend A comparative adhesion test in which the coated substrate is bent around a rod with a specified diameter. The deformed coating is observed visually and subjected to a tape test.

Adhesion test, breath An adhesion test that uses the internal stress in the film and the condensation of water from a person's breath, which enhances fracture propagation in a brittle material, to cause visual adhesion failure. Also called the **Mattox bad breath adhesion test.**

Adhesion test, indentation A comparative adhesion test where the surface is indented with a tip of a specific configuration and the fracture of the film around the indentation is observed visually.

Adhesion test, non-destructive A test that can be performed to establish the presence of a specified amount of adhesion without destroying the film. Examples: Tape-test of a mirror surface; Pull-to-limit wire-bond test.

Adhesion test, scratch An adhesion test whereby a loaded stylus with a specific tip configuration is pulled across the film surface under increasing load. The scratched surface is then observed visually for flaking and deadhesion and is correlated to the load at that point. During scratching **Acoustic emission** may also be monitored.

Adhesion test, stud-pull An adhesion test whereby a protrusion (stud) is bonded to the surface of the film and pulled in tension.

Adhesion test, tape A comparative **go** or **no-go** (pass or fail) adhesion test where an adhesive tape is applied to the surface of the film and pulled. If the film remains on the surface the adhesion is deemed good. May be used as a non-destructive adhesion test. The tape can be examined for **pull-outs**. See Non-destructive test.

Adhesion test, topple Where a bump is bonded to the film surface and pushed from the side until failure.

Adhesion test, wire-pull An adhesion test where a wire is bonded to the film surface, often by thermocompression bonding, then pulled until the wire breaks or the bond fails. The wire-bond test can be used in a non-destructive manner by pulling to a given pull, then using the wire in subsequent processing if the bond does not fail.

Adhesion test program A program designed to subject the film-substrate structure to the stresses (mechanical, chemical, thermal, fatigue) that it might see in subsequent manufacturing and service with adhesion testing to ensure the adhesion of the film under those conditions.

Adiabatic process A process where there is no gain or loss of heat to the surroundings.

Adsorbent The material doing the adsorbing.

Adsorbent capacity The amount of material the adsorbent can hold before becoming saturated. Example:

Grams of water per gram of Zeolite™.

Adsorption   Condition where material (adsorbate) is retained on the surface of the bulk (adsorbent). See Absorption.

Adsorption pump, vacuum (vacuum technology)   A capture-type vacuum pump that pumps by cryocondensation or cryotrapping on a surface whose temperature is less than -150°C. See Vacuum pump.

Aerosols (cleaning)   A suspension of very fine solid or liquid particles in a gas. The evaporation of the liquid aerosol can produce very fine particulate contamination if there is a residue.

Afterglow (plasma)   The region outside the plasma-generation region where long-lived plasma species persist. Also called **Downstream location**; **Remote location**.

Agglomeration (film growth)   Collecting into isolated regions (clumps).

Agile manufacturing   A modular manufacturing line organized such that the product can be changed easily. Example: From left-hand drive cars to right-hand drive cars.

Aging, natural   The change of property with time under normal conditions. See Accelerated aging.

Agitation (cleaning)   The introduction of turbulence into a fluid to enhance mixing and disrupt boundary layers near surfaces.

Air   The ambient gases that we breathe. Air contains gases, vapors and organic and inorganic particles.

Air, medical   Air that has been compressed and contains no substances, such as oil or carbon monoxide, that would be detrimental to a person's health. Also called **SCBA** (Self Contained Breathing Apparatus) **air**.

Air fire (cleaning)   Heating of a surface to a high temperature, in an air furnace or an oxidizing flame, to cause oxidation of contaminates. Example: Air fired alumina ceramics at 1000°C.

Air knife (cleaning)   A shaped jet of high-velocity air used to blow water from a surface as it passes in front of the air knife. See Drying.

Air shower (cleaning)   A downward flow of air used to blow particulates from the surface of clothing after donning cleanroom-type garments.

Alcohol (cleaning)   Any class of organic compounds containing an OH- group. Often used for wipe-down cleaning and drying.

Alcohol, anhydrous   An alcohol without water. Used as a wipe-down agent and to displace water from a surface.

Alcohol, denatured   Ethyl (grain) alcohol containing a material (denaturant) that makes it unfit to drink. Many materials used to denature alcohol will leave a residue on evaporation.

Aliphatic solvent (cleaning)   A type of solvent that consists of straight-chain hydrocarbons such as hexane and naphtha.

Alkaline cleaner (cleaning)   A basic cleaner that cleans by saponifying oils and chelating inorganic soils. The cleaner can also have agents for emulsifying, wetting, and penetrating; alkaline builders for neutralizing water hardness interference; corrosion inhibitors; etc. Alkaline cleaning is often followed by an acid rinse to neutralize the adhering alkaline material and remove non-soluble precipitates formed by reaction with the alkaline material. A **low alkalinity cleaner** has a pH of 7.5 to 9.0; a **strongly alkaline cleaner** will have a pH of 11.0 to 13.0.

Alloy   A mixture of two or more elements where there is mutual solubility such that the atoms are evenly dispersed among each other and the system is thermodynamically stable.

Alloy, pseudo   See **Mixture**.

Altered region (ion bombardment)   The region near the surface which has been altered by the physical penetration of the bombarding species or by "knock-on" lattice atoms. In the extreme case this can lead to the amorphorization of the region. See Near-surface region.

Alternating current (AC)   A potential that reverses polarity (and thus direction of current flow) each cycle.

Alternating ion plating (film deposition)   A repetitious process where a few monolayers of condensable film material are deposited, then the surface is bombarded followed periodically by more deposition and more bombardment. Also called **Pulsed ion plating**.

Alumina (substrate)   Aluminum oxide (Al_2O_3). Alumina substrates are usually in the form of fused material with some amount (4-15%) of silica glassy phase.

Aluminize   The process of depositing aluminum on a surface from a vapor.

Aluminize   The process of reacting a surface with aluminum to form an aluminum alloy or intermetallic phase.

Ambient conditions (vacuum technology, contamination control)   Conditions such as pressure, air composition, temperature, etc., that are present in the processing area.

Amine   Any one of a group of organic compounds derived from ammonia (NH_3) by replacement of one or more hydrogen atoms by organic radicals.

Ammonia (NH_3)   A chemical precursor vapor for nitrogen that is easier to decompose than is N_2.

Amorphous (crystallography)   Material with a grain size so small ($< 30\text{Å}$) that the X-ray diffraction pattern does not show any crystallinity. See Glassy.

Ampere (A)   Electrical current of one coulomb (1.6 X 10^{19} electrons) per second. Also called an **Amp**.

Amphoteric material   A material that can either gain

or lose an electron (i.e., act as either an acid or a base) in a chemical reaction. Example: Aluminum can form Al_2Cu or Al_2O_3.

Analytical electron microscopy (AEM) (characterization)   A combination of transmission electron microscopy (TEM) and electron diffraction.

Angle-of-incidence (film formation)   The angle of impingement of the depositing adatom flux as measured from the normal to the surface.

Angle-of-incidence effect (film growth)   The effect of angle-of-incidence of the adatoms on the development of a film morphology. See Columnar morphology.

Ångstrom (Å)   A unit of length equal to 10^{-10} meters or 0.1 nanometer.

Anhydrous (cleaning)   Without water. Example: Anhydrous (absolute) alcohol.

Anion (electroplating)   An ion that is negatively charged and will move toward the anode.

Anisotropy, film properties (film formation)   Properties that differ in different directions in the plane of the film. Often due to anisotropy in the flux of depositing material or anisotropy in the bombardment during deposition.

Annealing (glass)   Reducing the internal strain by raising its temperature to the point (**Strain point**) that atoms can move so as to relieve the strain or other thermodynamic differences.

Annealing (metal)   Reducing the internal strain by raising its temperature (**Annealing temperature, Recrystallization temperature**) to the point that atoms can move so as to relieve the strain or other thermodynamic differences. Annealing results in softening of the metal.

Anode   The positive electrode in a gas discharge or electroplating bath.

Anode-to-cathode ratio (electroplating)   The ratio of the surface area of the anode to that of the cathode.

Anodic arc, plasma (plasma technology)   An arc vaporization source where the vaporized material originates from a molten anode electrode. Also called a **Distributed Arc**. See Arc source.

Anodic cleaning (cleaning)   Cleaning a surface by removing (off-plating) material from the anode in an electrolytic cell. Also called **Electrolytic cleaning**. Called **Electrolytic pickling** if the solution is acidic.

Anodic etching   Roughening or exposing grain structure by anodic dissolution (off-plating) in an electrolytic cell.

Anodization   The electrolytic conversion of an anodic surface in an electrolysis cell or oxygen plasma (**Plasma anodization**) to an oxide. Example: Aluminum anodization.

Anodize, barrier   A non-porous anodic oxide that can be formed on materials such as aluminum, titanium, and niobium. The thickness of the oxide is proportional to the anodizing voltage applied.

Anodize, porous   A porous anodic oxide that is formed in an electrolytic bath that corrodes the oxide as it is being formed, thus giving porosity in the oxide and allowing a thick oxide layer to be formed. Generally the porous coating is sealed (expanded) by hydration in a hot water bath.

Antiferromagnetic   A material in which the electron spins are ordered in an antiparallel arrangement such that there is zero magnetic moment. Example: Cr.

Antioxidant   A substance added to a plastic to slow the degradation by oxidation.

Antireflection (AR) coating (ARC) (optics)   A film structure designed to reduced reflection over a region of the spectrum so that radiation in that spectral region is transmitted into the substrate.

Antiseize compounds (vacuum technology)   Material applied to a surface to prevent cold welding and galling. Example: Silver-plated stainless steel bolts. See Lubricant, vacuum.

Antistatic agent   Chemical substances that increase the surface conductivity of plastic materials and are used to prevent surface charge buildup. Often they are ionic materials that absorb water to become conductive.

Applied bias (PVD technology)   An electrical potential applied from an external source. See Bias.

Aqua regia (cleaning)   A mixture of hydrochloric acid and nitric acid in a ratio of 3 to 1.

Aqueous cleaning   Water-based cleaning such as mixtures of water, detergents, and other additives that promote the removal of contaminants.

Aqueous solution   A solution where water is the solvent.

Arc   A high-current, low-voltage electrical discharge between two electrodes or between areas at different potentials. See Arc source.

Arc, gaseous   An arc formed in a chamber containing enough gaseous species to aid in establishing and maintaining an electrical arc. See Arc, vacuum; Flashover.

Arc, vacuum   An arc formed in a vacuum such that all of the ionized species originate from the arc electrodes. See Arc, gaseous.

Arc cleaning (plasma spraying, cleaning)   The use of a cathodic arc to clean and etch (roughen) a surface prior to deposition.

Arc source, anodic arc   An arc vaporization source where the vaporized material originates from the anode surface, which is liquid. Also called a **Distributed Arc Source**.

Arc source, cathodic arc   An arc vaporization source where the vaporized material originates from the cathode surface, which is usually solid.

Arc source, filtered An arc vaporization source designed to filter out the macros, generally by deflecting the plasma. See Plasma duct; Macros.

Arc source, random arc Cathodic arc where the arc is allowed to move randomly over the cathode surface.

Arc source, steered arc A cathodic arc where the arc is moved over the surface under the influence of a magnetic field.

Arc suppression Techniques for quenching an arc before it becomes too destructive. These include: shutting off the power or introducing a voltage pulse with an opposite polarity.

Arc vapor deposition (physical vapor deposition, vacuum deposition processes) Film deposition process where the source of vapor is from arc vaporization.

Arc vaporization Vaporization of a solid (cathodic) or liquid (anodic) electrode material using a vacuum or gaseous arc. Characterized by high ionization of the vaporized material. Also called **Arc evaporation**.

Arc-wire spray A thermal spray process where the tip of a wire(s) is melted in an electric arc and the molten material is propelled to the substrate by a gas jet.

Architectural glass Glass used in buildings. Usually windows.

Archival samples Samples retained after processing has been performed to allow comparison with material at a later stage or after being placed in service. See Control samples; Shelf samples.

Argon (sputtering) An inert (noble) gas used for sputtering because it is relatively inexpensive compared to other inert gases and has a reasonably high mass (40 amu).

Aromatic solvents (cleaning, topcoats, basecoats) Solvents based on benzene-ring molecules such as benzene, xylene, and toluene. Used as diluents in acrylic lacquers.

Arrhenius equation A equation relating a rate, such as a chemical reaction rate, to an activation energy and the temperature.

Art (*Archaic*) An old term for technology. Example: Prior art.

As-received material (manufacturing) Material that first enters the processing sequence. The material may be from an outside supplier or from a previous processing sequence. See Process Flow Diagram; Inspection, incoming.

ASA flange (vacuum technology) A flange for joining tubing that has a specific bolt pattern for each diameter.

Ashing (cleaning) Reducing a material to non-volatile residues (ash) by high temperature or plasma oxidation.

ASME Boiler and Pressure Vessel Code (vacuum technology) The American Society of Mechanical Engineers code by which the material, material thickness, design, and construction methods are specified for pressure vessels. Since a vacuum chamber is a pressure vessel, the code is often used in specifying the construction of vacuum chambers.

Aspect ratio (surface, semiconductor) The ratio of the depth to the width of a feature such as a via (hole) or trench in a surface. Example: The aspect ratio of a **via** could be 4 : 1.

Asperity (surface) A small protuberance from a surface. It may be of the bulk material or be an inclusion.

Asymmetrical AC Where the amplitude, duration, and/or waveform of the voltage in one polarity of an alternating current (AC) voltage cycle is different from that in the other polarity. Also called **Unbalanced AC**. See Alternating current (AC).

Atom The basic unit of a chemical.

Atomic force microscope (AFM) (characterization) A stylus surface profilometer that measures the deflection of a probe mounted on a cantilever beam. The AFM can be operated in three modes: contact, non-contact and "tapping" mode. Also called the **Scanning force microscope (SFM)**.

Atomic layer deposition (ALD) A technique of pulsed chemical vapor deposition (CVD) that deposits small amounts of material periodically and is used to fill high-aspect-ratio surface features and porous materials.

Atomic mass unit (amu) The atomic mass unit is defined as $\frac{1}{2}$ of the mass of the ^{12}C isotope. Also called the **Unified atomic mass unit (u)**. One amu = 1.66 x 10^{-24} g.

Atomic peening (film formation) The continuous or periodic bombardment of a depositing film with high energy atoms or ions to densify the depositing film material. Atomic peening tends to introduce compressive stress into a surface or growing film.

Atomic percent (alloy) The percentage by atomic ratio of one material in an alloy composition. Abbreviated at%. See **Weight percent**. Example: An alloy of W:30at%Ti has the same composition as W:10wt%Ti.

Atomically clean surface A surface that does not contain an appreciable fraction of a monolayer of foreign material on the surface. Very difficult to obtain and retain.

Auger electron emission The emission of electrons from an excited atom that have a characteristic energy due to specific transition between orbital states in the atom.

Auger electron spectroscopy (AES) (characterization) A surface analytical spectroscopy technique that uses energetic electrons as the probing species and Auger electrons as the detected species.

Augmented plasma (plasma technology) A plasma that has had electrons injected from an outside source to enhance ionization.

Aurora Borealis coating (decorative coating)   Coating with a rainbow of colors formed by depositing films or anodizing surfaces to give colored interference patterns.

Autocatalytic plating   Deposition of a coating from a solution by use of a reducing agent in the solution rather than an external-applied electrical potential. Also called **Electroless deposition, Autodeposition, Autophoretic deposition.**

Automotive glass   Glass used in the windows of automobiles. Often bent into a curved form. Usually tempered or laminated safety glass.

Auxiliary plasmas (plasma technology)   A plasma established in a processing system to assist in some aspect of the processing separate from the main processing event. Examples: Plasma cleaning in a vacuum deposition system; Plasma activation of the reactive gas near the substrate in a reactive magnetron sputter deposition system.

Availability, reactive gas (film formation)   The availability of the reactive gas over the surface of the film being deposited. Since the surface of the film is continually being buried, reactive gas availability is an important parameter in reactive deposition.

Avogadro's Number   The number of molecules contained in one mole (gram-molecular-weight) of a substance. The value is 6.023×10^{23}.

Avoirdupois (a) weight system   Common pound and ounce system where 1 ounce (oz) (a) = 28.4 grams and 1 pound (a) = 16 oz (a). See Troy weight system.

Azeotropic mixture (cleaning)   Solvent mixture where the vapor has the same composition as the liquid.

Back-diffusion (vacuum technology)   Flow of vapor in a direction opposite to that of the flow of gas being pumped. Occurs in the molecular flow range. Also called **Backstreaming**.

Back-end (semiconductor technology)   Final processing such as dicing, wire bonding, encapsulation, test, assembly packaging, etc. See Front end.

Back-scattering   Scattering of particles in a direction counter to that of the main particle flow.

Backcoat   The protective coating that is applied to the film on the second surface. See Second surface coating. Example: Polymer coating applied to the aluminum reflector coating on a back surface mirror.

Backfill (vacuum technology)   Raising the system pressure with a specific gas. Example: Backfill with dry gas and raising the pressure in order to establish a plasma. See Venting.

Backing plate (sputtering target)   The plate that the target material is bonded to that allows mounting to the cooling portion of the sputtering target assembly.

Backing pump (vacuum technology)   A vacuum pump used to keep the discharge pressure of a high vacuum pump below some critical value. The backing pump may be also used as a roughing pump. Also called a **Forepump.**

Backpressure (vacuum technology)   The pressure in an exhaust system that impedes the flow of gas through the exhaust system.

Backside film (semiconductor processing)   Film or coating deposited on the backside of a silicon wafer during processing of the frontside (the side on which the device structure is being built).

Backstreaming (vacuum technology)   Movement of gases or vapors from the high pressure to the low pressure region of a vacuum system. Also called **Back-diffusion.**

Baffle (vacuum technology)   A system of surfaces designed to minimize backstreaming either by condensation or reflection. Also called a **Trap**.

Baffle (PVD technology)   A system of surfaces to prevent a cold vacuum pumping surface from seeing the thermal radiation from the processing chamber.

Baffle source (evaporation, PVD technology)   An evaporation source in which the vapor must collide with several hot surfaces before it can leave the source. Used to evaporate materials such as selenium and silicon monoxide, which vaporize as clusters of atoms or molecules.

Bag filter (vacuum technology)   Mechanical filter to prevent particulates from entering the vacuum pumping system.

Bag-check (vacuum technology)   Covering a vacuum system with a bag filled with helium to measure the total real leak rate into the system. Also called a **Hood test**.

Bake-out (vacuum technology)   The heating of a vacuum system to a high temperature (i.e., 400°C) to accelerate outgassing and desorption from materials and surfaces in the vacuum system.

Baking, vacuum (cleaning)   Heating of a material at an elevated temperature for a period of time sufficient to reduce volatile constituents such as water, solvents, and plasticizers to an acceptable level. Care must be taken not to heat the material to a temperature at which it will decompose. The necessary time and temperature is generally determined using weight-loss or mass spectroscopic analysis.

Baking soda (cleaning)   Sodium bicarbonate. Used as a water-soluble mild abrasive.

Ball bond A wire bond to a film consisting of a ball formed on the tip of a wire that is bonded to the surface under heat and pressure (**Thermocompression bonding**) or under pressure and ultrasonic scrubbing (**Ultrasonic bonding**). See Wire bond.

Ballast orifice (vacuum technology) An orifice upstream of the mechanical pump that can be used to allow dilution of the pumped gas with dry gas to ensure that vapors in the pumped gas do not condense during compression in the mechanical pump. The ballast orifice also allows the foreline portion of the vacuum pumping manifold to return to ambient pressure in case the mechanical pump stops because of a power failure or a broken belt. This avoids **suck-back**.

Ballast tank (vacuum technology) A large volume that can be continuously pumped and is used to assist in rapid roughing by opening the much smaller volume of the deposition chamber to a ballast tank for the initial rough pumping.

Ballast valve (vacuum technology) A valve in or just before the mechanical pump that can be used to allow dilution of the pumped gas with dry gas to ensure that vapors in the pumped gas do not condense during compression in the mechanical pump. The ballast valve can also be opened automatically to allow the foreline portion of the vacuum pumping manifold to return to ambient pressure in case the mechanical pump stops because of a power failure. This avoids **suck-back**. See Suck-back.

Balloon gasket (vacuum technology) An inflatable elastomer gasket used to seal non-parallel sealing surfaces.

Band-pass filters (optical coatings) Optical coatings that allow a band of specific wavelengths to pass through and others to be reflected or absorbed. See Heat mirror; Dichroic coatings.

Banding (PVD technology) A striped pattern on large-area substrates or webs due to variation in film thickness, morphology, or composition across the width of the web.

Bar (pressure) Pressure equal to 10^5 Pascals. 1 bar = 0.98692 atmospheres = 750.06 Torr. Pressure unit commonly used in Europe. A **Millibar** is 0.001 bar.

Barrel plating (electroplating, PVD technology) Plating objects that are loose inside a rotating grid structure (cage or barrel) so that they are tumbled and completely covered. See Fixture.

Barrier film (diffusion, permeation) A film used to reduce the diffusion into a surface or through a film. Examples: TiN underneath aluminum metallization on silicon to prevent diffusion of Al into the silicon on heating; Aluminum film on a polymer web to reduce water permeation through packaging material.

Base Any chemical species capable of accepting a proton (hydrogen ion) from another species. (Example: OH-). An acid yields hydrogen ions (H+) by reaction with the solvent while a base forms hydroxyl ions, OH-. See Acid.

Base pressure (vacuum technology) A specified pressure for the system to begin the next sequence in the processing. See Pumpdown time; Ultimate pressure.

Basecoat (PVD technology) A film, often a polymer, that is applied to a surface to produce a smooth surface (**Flow coat**), to seal in material that will outgas during vacuum processing, or to provide a "**Glue-layer**" for adhesion.

Baseplate The large-area stationary surface, usually horizontal, on which a moveable vacuum chamber seals and that contains many of the feedthroughs into the system. See Collar.

Basic surface (film formation, adhesion) A surface capable of supplying an electron to an atom on its surface. See Acidic surface.

Batch (PVD technology) A group of substrates that are processed in the same fixture in one "run."

Batch processing system A system where the processing chamber is opened to the ambient each time the fixture is placed into or removed from the chamber. Also called a **Direct-load system** (preferred).

Bayard-Alpert gauge (vacuum technology) A hot cathode ionization gauge using a fine-wire ion collector to minimize X-ray effects in the gauge.

Bead blasting, glass (cleaning) Subjecting a surface to bombardment by beads (usually glass) entrained in a high-velocity gas flow to abrasively clean the surface.

Beam density Particle flux (particles per cm^2) in the beam.

Beam intensity Power density of the beam (watts per cm^2).

Beam neutralization The addition of electrons to an ion beam so that there is no net charge in a volume of the beam, even though individual species in the beam still have an electrical charge.

Beam splitter (optics) An optical filter or reflector that reflects some of the incident radiation and transmits the rest. Also called a **Beam divider**.

Bell jar (vacuum technology) A moveable glass or metal vacuum chamber that is generally cylindrical with a domed top that seals to a baseplate using an elastomer seal. Most often removed by lifting from the baseplate.

Bellows, metal (vacuum technology) An axially expandable tube of metal that is used to allow alignment of flanges, isolation from vibration, or motion in a linear direction.

Belt furnace Furnace where the part is moved through the hot zone on a moving belt. This allows a controlled heating rate, time-at-temperature, and cooling rate.

Bend test (adhesion)   An adhesion test where the coated substrate is bent around a radius and the coating is observed for spallation from the substrate. See Adhesion test.

Beta backscatter (thickness measurement)   Beta particles (electrons) from a radionuclide source are scattered from a film on a surface into a geiger counter. By calibration with known film thickness the signal from the counter can be used to measure the film thickness. Various radionuclides can be used to give beta particles with different energies.

Beta particles   Electrons from radioactive sources.

Beta test (semiconductor processing)   Evaluation of equipment by an OEM (**original equipment manufacturer**) under production conditions to determine what changes should be made before supplying the final version of the equipment to the user.

Bias (statistics)   A systematic error that contributes to the difference between the mean of the measurement and an accepted reference or true value.

Bias, applied (PVD technology)   An electrical potential applied from an external source.

Bias, magnetic (PVD technology)   Magnetic field in the vicinity of the substrate during deposition to affect the structure and orientation of deposited magnetic films.

Bias, self (plasma technology)   An electrical potential on a surface generated by the accumulation of excess electrons (**Negative self-bias**) or positive ions (**Positive self-bias**). See Sheath potential.

Bias sputtering   Sputter deposition with a bias on the substrate to accelerate ions to the surface during deposition. See Ion plating.

Biaxial orientation (BO) (substrate, polymer web)   The process of stretching a plastic film (usually at elevated temperatures) in both the machine and transverse directions so as to achieve similar tensile, modulus, and elongation properties in the film.

Binding energy   The strength of the chemical bond between atoms.

Bipolar DC (poor terminology)   A term used by some to describe a potential that reverses polarity during some part of each cycle (thus is really an AC potential). See AC potential; Asymmetrical AC.

Bipolar pulse power (plasma technology)   Applying electrical power with a periodic waveform with either an off-portion of the waveform on each cycle or a portion of the waveform having an opposite polarity on each cycle. See Mid-frequency.

Bit (semiconductor)   A unit of information represented by a change of state (i.e., on then off). See Byte.

Bit density (semiconductor)   The number of bits (in-

formation storage) per unit area on a silicon chip (or magnetic tape).

Black body (radiation)   A surface that absorbs all radiation of any wavelength that falls on it. The surface will have an emittance of unity.

Black body radiation   The characteristic radiation from a black body surface at a specific temperature.

Black breath test (cleaning)   Condensation of moisture from a person's breath on a cleaned surface. Uniform nucleation indicates a uniformly clean surface (if the contamination is not hydrophilic).

Black out   When the power line voltage goes to zero. See Brown out.

Black sooty crap (BSC)   Ultrafine particles formed by vapor phase nucleation in a gaseous environment. See Soot; Ultrafine particles.

Blank off (vacuum)   To place a solid plate over the opening of vacuum plumbing. Example: Blank off pressure (vacuum) of a pump.

Blanket metallization (PVD technology)   Metallization over the whole surface. See Selective metallization.

Bleb (glass)   A bump on the surface of glass caused by a bubble or an inclusion in the glass.

Bleed (vacuum technology)   The continuous admission of a small amount of gas into a vacuum or plasma system.

Blister (adhesion)   An enclosed separation of a coating from the substrate.

Blocking (web coating)   When the film sticks to itself in the wound condition on the roll.

Blocking capacitor   A capacitor that is place in an rf circuit to retain some of the electrons and reduce the number of electrons (voltage) that appears on the rf electrode.

Bloom (float glass)   Bluish haze on float glass caused by wrinkling of the surface as the glass surface adsorbs oxygen and expands.

Blooming (*Archaic*)   A British term for depositing a thin film by vacuum evaporation. Example: Double blooming is to deposit a two-layer film.

Blow hole (basecoat, topcoat)   A void in a flow coating formed by outgassing during heating before the coating is cured.

Blow-off (cleaning)   A method of cleaning particulates from a surface using a high-velocity stream of clean gas. When blowing-off the surface of an insulator the gas should be ionized to prevent static charge buildup on the insulating surface.

Blower (vacuum technology)   A low-compression mechanical, compression-type vacuum pump. Example: Roots blower.

Boat source (evaporation)   An evaporation source where

the charge is contained in a cavity in a surface. Generally the boat is of tungsten, tantalum, or molybdenum and is heated resistively. The cavity may be coated with a ceramic so that the molten charge does not come into contact with the metal. See Evaporation source.

Body covering (cleaning)   The coat, head covering, face covering, shoe covering, gloves, etc. used to contain particulate contamination generated by a person's body and clothes.

Body tinted glass   Glass that is colored due to incorporation of a coloring agent in the glass.

Boiling point   When the vapor pressure of the material is the same as the ambient pressure. Example: At sea level the boiling point of water is 100°C.

Boiling beads (evaporation)   Solid masses added to a liquid to prevent splattering and spitting during boiling or evaporation. Example: Tantalum shot in molten gold to prevent spitting by vapor bubbles rising through the molten gold.

Boltzmann's constant (k)   The ratio of the Universal Gas Constant to Avogadro's number. The constant (k) in the equation $E = 3/2\ kT$ which gives the mean energy (E) of a free particle at a temperature T (K). $k = 1.38 \times 10^{-16}$ erg/deg (K).

Bombardment-enhanced chemical reactions (film formation)   Chemical reactions on a surface that are enhanced by bombardment by high-energy atomic-sized particles. The effect is due to heating, dissociation of adsorbed species, production of electrons, etc. Important effect in reactive deposition, PECVD, plasma etching, and reactive ion etching.

Bombing (leak detection)   To place a container in a high-pressure gaseous environment (usually helium) to force the gas through leaks into the interior.

Bond energy   The energy released by the formation of a molecule from its constituent atoms. Individual bond energy is calculated by dividing the dissociation energy by the **coordination number**. Also called **Bond strength**. For example: Bond strengths for oxide glass components are: $SiO_2 = 106$ kcal/mole, $Al_2O_3 = 101$ to 79 kcal/mole (depending on the coordination number), $PbO = 36$ kcal/mole, and $Na_2O = 20$ kcal/mole.

Bondability (semiconductor processing)   The ease with which a wire can be attached to the surface.

Bonding (sputtering target)   The attachment of the sputtering target to the backing plate using a technique that gives good thermal contact.

Bonding pad   An area of film where a contact such as a wire is to be bonded, usually under heat and pressure (Example: Thermocompression bond). Substrate under the film is often put under significant stress during the bonding operation.

Book-to-bill ratio (business)   The ratio of orders received to orders shipped in a particular month.

Booster pump (vacuum technology)   A pump used between the high vacuum pump (particularly the diffusion pump) and the backing pump in order to increase the throughput in the medium vacuum range and decrease the volumetric flow through the backing pump. Example: Diffusion pump exhausts into a Roots blower (booster pump) then into an oil-sealed mechanical pump. See Vacuum pump.

Booties (contamination control)   Shoe coverings used in a cleanroom.

Boronize (substrate)   The process of diffusing boron into a surface region containing Mo, Cr, Ti, etc., so as to form a surface layer (**Case**) containing boride compound particles dispersed through the layer.

Boundary layer (cleaning)   The layer of stagnant fluid next to a surface through which cleaners must diffuse to reach the surface. See Agitation.

Boundary layer (electroplating)   The layer of stagnant fluid next to a surface through which ions must diffuse to reach the surface. See Agitation.

Box coater (deposition chamber)   A direct-load deposition chamber in the form of a flat-sided box, often with gussets, with one or more sides being a door. See Deposition system.

Boyle's Law   For an ideal gas at a fixed temperature the product of the volume of the gas and its pressure is equal to a constant.

Brass   A copper zinc alloy (Cu : 5-40% Zn).

Braze alloy (vacuum technology)   A metallic alloy that melts above about 450°C and is used to join two materials together.

Bright dip (surface)   A chemical treatment that tends to preferentially etch the high points on a surface, thus increasing the smoothness of the surface. Example: 10% HCl on aluminum.

Brightness   One component of color. The component of color that gives the perception of intensity. Also called **Luminance**. See Color.

Brittle fracture (adhesion)   Fracture of a material with little or no plastic deformation.

Brittle material   A material that allows little or no plastic deformation before failure. Generally such a material has a low fracture toughness.

Bronze   A copper tin alloy (Cu : 1-20% Sn) that has many of the same machining properties as brass but is more expensive. A typical bronze is bell-bronze (77% copper, 23% tin).

Brown-out   When the power line voltage drops below a specific voltage but is still greater than zero. A brown-out can affect the operation of electrical gear such as motors, electronics, etc. See Black-out.

Brush plating (electroplating)   Plating where the anode is a moveable electrode and the electrolyte is held in an absorbent material (swab) on the anode. The part to be coated is made the cathode.

Bubbler (agitation)   Perforated pipe distributor for fluids or gases used in the bottom of fluid tanks for agitation. Also called a **Sparger**.

Buckles (web coating)   Ridges of film that extend across the roll or around the roll of film material.

Buffer layer (cleaning, etching)   A layer of material that has properties or crystal structure, intermediate between the film and the substrate materials, and allows gradation of properties between the two materials. See Compliant layer.

Buffered solutions (cleaning)   A chemical solution formulated to minimize the change of hydrogen ion concentration in the solution due to chemical reactions.

Bulk getter (vacuum technology)   A mass of material that retains gases that diffuse into it. See Getter.

Bulkhead mounting (vacuum technology)   When a chamber is mounted through a wall such that the chamber opening is on one side and the pumping plumbing is on the other side of the wall. This design ensures that persons working on the pumping system do not contaminate the processing environment of the opening side. See Pass box.

Bunny suit (cleaning)   Body covering that covers the head, neck, torso, legs, and feet.

Burnishing   Smearing a soft metal either by mechanical contact with a smooth surface such as steel balls, or by the use of a mild abrasive. Examples: **Barrel burnishing**; **Vibratory burnishing.**

Burping (vacuum pump)   The sudden release of gas from a vacuum pump.

Burr   A thin protruding piece of metal along an edge that is left after a forming process.

Byte (b) (semiconductor)   An association of binary bits that act as a unit in a computer.

C

Calcium carbonate (CaCO₃) (cleaning)   Used as a polishing/cleaning abrasive. Insoluble in water, soluble in acids. Also called **Chalk**.

Calibrated leak   A leak that has a known leak-rate (Torr-liters/sec) for a specific gas under specific conditions. Used to calibrate leak detectors.

Calibration   To determine by comparison to a standard the absolute value of each scale reading of a sensor device. Comparison must be done in a specified manner

under specified conditions. See Standards, primary; Standards, secondary.

Calibration log   The document describing when a unit was calibrated, by what method, and the name of the person who did the calibration.

Canted spring seal (vacuum technology)   A slit tubular seal that has the restoring force provided by a canted coil spring inside the tube.

Capacitance manometer (vacuum technology)   A vacuum gauge that uses the deflection of a diaphragm, as measured by the changing capacitance (distance) between surfaces, as an indicator of the pressure differential across the diaphragm, the pressure on one side being a known value. See Vacuum gauge.

Capacity, pump (vacuum technology)   The amount of a specific gas that a capture pump, such as a cryopump, can contain and still pump effectively. When this value is exceeded the pump is ineffective and must be regenerated. See Regeneration.

Capillary action   The combination of adhesion and cohesion that causes fluids to flow or rise between closely spaced surfaces.

Capillary waves (substrate)   Periodic waviness on a polished surface. See Orange peel.

Capture pump (vacuum technology)   A vacuum pump that captures and holds the gases and vapors being pumped. See Vacuum pump.

Carbides, metal (corrosion)   Carbon-metal compounds that can be formed in some alloys in the **Heat affected zone (HAZ)**, during welding, that can give galvanic corrosion problems. See Stainless steel; Low carbon steel.

Carbon dioxide (CO₂), liquid (cleaning)   Liquefied carbon dioxide used as a solvent. See Green cleaning.

Carbon dioxide (CO₂), snow (cleaning)   Solid carbon dioxide that is used to abrasively clean a surface and is formed by expansion and cooling of a jet of compressed carbon dioxide gas.

Carbonitriding (substrate)   Hardening by diffusion of both carbon and nitrogen into a metal surface to form both carbide and nitride phases dispersed in the surface region. See Carburizing; Nitriding.

Carbonyl (carbonyl group)   The radical (C=O). Example: $Mo(CO)_6$.

Carboxyl (carboxyl group)   The (COOH) group.

Carburizing (substrate)   The process of diffusing carbon into a surface region of an alloy containing Cr, Ni, Mo to form a carbide phase and give dispersion strengthening.

Carcinogenic (chemical)   A chemical that has been shown to cause cancer in mice. See Mutagenic.

Carrier gas (CVD)   Gas used to decrease the concentration of reactive gases in CVD reactions without changing the total pressure or to entrain and carry vapors into the reaction chamber. Also called a **Diluent gas**.

Carryover (cleaning)   Water or chemicals that are carried from one tank to another and must be replenished by using **Makeup** water or chemicals.

Cascade rinse (cleaning)   Rinsing using a series of containers (tanks) having increasingly pure water. Water flows over the lip of one container into the next container having lower purity water. The surface being rinsed goes from the lower purity to the higher purity rinse tank. Also called **Counterflow rinse**. See Spray rinse.

Case (substrate)   A hardened surface region that can extend many microns into the surface.

Case hardening   Surface hardening by forming a dispersion-strengthened surface layer (case) of appreciable depth by one of several techniques.

Catalyzed reaction   A chemical reaction whose rate is increased by a material that is not consumed in the reaction.

Cathode   The negative electrode in a gas discharge or an electroplating bath.

Cathode spot (plasma technology)   The area on the cathode, under normal glow discharge conditions, in which the current is concentrated. As the current increases the spot becomes bigger in order to maintain a constant current density in the cathode spot. In **Abnormal glow discharge** the cathode spot covers the whole cathode area.

Cathodic arc (PVD technology)   A vaporization source where the vaporized material originates from a high current density arc on the cathode surface, which is usually solid. See Anodic arc.

Cathodic cleaning   Cleaning in an electrolytic cell where the surface to be cleaned is the cathode. See Anodic cleaning.

Cation (electroplating)   An ion that is positively charged and will move toward the cathode. See Anion.

Cationic detergent (cleaning)   A detergent that produces aggregates of positively charged particles with colloidal properties.

Cavitation (cleaning)   Formation of vapor-filled voids (bubbles) in a fluid under tensile stress. The voids grow to a size determined by the surface tension of the fluid, then collapse. If the voids are in contact with a surface the collapse produces a jet of fluid that can clean the surface and cause cavitation erosion of the surface. See Ultrasonic cleaning.

Ceiling (safety)   The exposure limit to which a worker must not be exposed to even instantaneously, as set by OSHA. See Threshold limit; Time-weighted average; Short-term exposure limit.

Centigrade temperature scale   A temperature scale in which the freezing point of water is taken as 0°C and the boiling point of water, under standard pressure, is taken as 100°C. Also called the **Celsius temperature scale**. See Temperature scale.

Cerium oxide (CeO$_2$)   Fine polishing compound used to polish glass.

Chain clamp (sealing)   A flexible chain, resembling a bicycle chain, that holds two tapered mating flanges in its links and when tightened applies a clamping-force at many contacting points.

Chalk (cleaning)   Calcium carbonate (CaCO$_3$). Used as a polishing/cleaning abrasive. Insoluble in water, soluble in acids.

Chamber, deposition   See Deposition system.

Channeling (ion bombardment)   The preferential movement of an energetic ion or atom along the open region between crystallographic planes in a solid crystal.

Characterization, extensive   Determining some film properties, such as crystallography, gas content, chemical concentration gradient, etc., which will take a significant period of time.

Characterization, film   Determining the properties of a film using specified characterization techniques.

Characterization, first check   Determining some film properties such as color, after the fixture has returned to atmospheric pressure but before the substrates have been removed from the fixture. See Position equivalency.

Characterization, functional   Characterization of the properties of the film that can or will be used in the final product. Example: Optical reflection.

Characterization, in situ   Determination of some film properties, such as thickness, optical properties, etc., during the deposition process or before the system has been returned to atmospheric pressure.

Characterization, non-destructive   Determination of some film properties, such as thickness, optical properties, etc., without affecting the film in a detrimental manner.

Characterization, rapid feedback   Determining some film properties such as sheet resistivity, thickness, or chemical composition, soon after the substrates have been removed from the fixture. See Position equivalency.

Charcoal, activated (vacuum technology)   See Activated carbon.

Charge (evaporation)   The material to be vaporized that is placed in a thermal vaporization source.

Charge exchange (plasma)   When a positive ion gains an electron from a neutral atom. If the ion has a high energy, the process produces a high-energy neutral and a low-energy ion.

Charge site   An immobile lattice site that has a deficiency or excess of charges.

Charge separation   When two atoms, molecules, or

surfaces are separated and one material has excess electrons and the other has a deficiency of electrons. This situation can cause arcing. See Exoemission.

Charging, hydrogen (cleaning)   When hydrogen is introduced into a surface by a chemical, electrochemical, or implantation action so as to form a high chemical gradient between the surface region and the bulk of the material. Examples: Electroplating of chromium introduces large amounts of hydrogen into the chromium; Acid cleaning of some metal surfaces introduces hydrogen into the surface.

Charles' Law   For an ideal gas at a constant pressure the volume of a fixed mass of gas varies directly with the absolute temperature.

Chelating agents (cleaning)   An organic compound that reacts with metal ions in solution and prevents them from reacting with other ions and being precipitated as an insoluble compound. Can pose a water pollution problem. Examples: Chelating agents include ethylene diamine tetraacetic acid (EDTA); Amine compounds.

Chemical bond   The strong attractive forces that exist between atoms or molecules due to electrical effects within and between atoms and molecules.

Chemical bonding, covalent   The chemical bond that is formed between two atoms in which each contributes one electron. If the electrons are shared unequally it is a **Covalent polar bond**. Also called **Electron pair bond**.

Chemical bonding, ionic   The chemical bond that is formed between atoms that have opposite electrical charges due to the transfer of an electron from one to the other. Example: NaCl.

Chemical bonding, metallic   The chemical bond that results from the immersion of the metallic ions in a "continuum" of freely moving electrons.

Chemical bonding, polar   The chemical bond that results between two atoms or molecules that are oppositely polarized.

Chemical bonding, Van der Waals   The chemical bond that results from the dipole interaction between two atoms or non-polar molecules. Also called **Dispersion bonding.**

Chemical conversion   The formation of a surface layer due to chemical reaction with a selected material. Examples: Chromate conversion; Phosphate conversion.

Chemical deposition   The deposition of a metal film by precipitation where another metal ion displaces the depositing atom in a solution of the metal salt. Example: Chemical silvering.

Chemical equivalent weight   Gram atomic (molecular) weight divided by the valence of the ion. Also called Gram equivalent weight. See Mole.

Chemical etch-rate test (characterization)   The rate (Ångstroms per minute or mass per unit area per minute) at which material is removed by chemical etching.

Chemical etching (cleaning)   The removal of material by chemical reaction with a fluid (**Wet chemical etching**) or vapor (**Vapor etching**) to produce a soluble or volatile reaction product. The etch rate is affected by the density, porosity, and composition of the film.

Chemical hoods   Enclosed, ventilated (air flow > 100 ft/min) region for performing chemical processes and isolating the processes from other processes.

Chemical polishing   Chemical removal of the high points on a surface.

Chemical potential   The chemical concentration difference between two regions.

Chemical pumping   The removal of gas by having it react with a material to form a compound having a low vapor pressure. Also called **Gettering**. See Getter pumping; Getters; Ion pumping.

Chemical roughening   Surface roughening by the preferential attack of features such as crystallographic planes, grain boundaries, and lattice defects.

Chemical silvering   The deposition of silver from solution by the reduction of a silver-containing chemical. Example: Used in coating backsurface mirrors and vacuum flasks.

Chemical solution, strength of   See Normal solution; Molality; Percent solution; Specific gravity.

Chemical sputtering (cleaning, etching)   Bombardment of a surface with a chemical species (e.g., Cl, F) that forms a volatile compound with the surface material. See Reactive plasma cleaning; Reactive plasma etching; Physical sputtering.

Chemical strengthening, glass (substrate)   Placing the surface of the glass in compression by replacing small ions (e.g., Na) with larger ions (e.g., K) in the surface region by diffusion.

Chemical vapor deposition (CVD)   The deposition of atoms or molecules by the reduction or decomposition of a chemical vapor species (precursor vapor) that contains the material to be deposited. Example: Silicon (Si) from silane (SiH_4). See Vapor Phase Epitaxy; Decomposition reaction (CVD); Reduction reaction (CVD); Disproportionation reaction (CVD).

Chemical vapor precursor (CVD, reactive deposition)   A gaseous chemical species that contains the species to be deposited. Examples: Silane (SiH_4) for silicon; Methane (CH_4) for carbon.

Chemical-mechanical cleaning (cleaning)   Combining chemical etching with mechanical abrasion.

Chemical-mechanical-polishing (CMP) (semiconductor processing)   A combination of chemical polishing and mechanical polishing that is used to planarize a surface.

Chemisorption The retaining of a species on a surface by the formation of strong chemical bonds (> 0.2 eV) between the adsorbate and the adsorbing material. See Physisorption.

Chill drum (web coating) The cold drum in contact with the web during the actual film deposition part of the web coating process to remove heat from the web. See Free-span.

Chip (electronic) A discrete device such as a transistor, a capacitor, a resistor, etc., on a substrate such as silicon or ceramic.

Chip (flaw) A region of a brittle material that is missing due to fracture, usually due to handling. The chip can be an **Edge chip** or a **Surface chip**.

Chip (semiconductor) One of many discrete semiconductor devices on a silicon wafer. As fabricated, each wafer contains many chips and is "diced" to create individual chips.

Chlorinated solvents (cleaning) Solvents containing carbon and chlorine such as **Trichloroethylene (TCE), Methylene chloride (MEC), Perchloroethylene (PCE)** and **1,1,1 trichloroethane (TCA)**. Very effective solvents but regulated because of health and environmental concerns. Example: Carbon tetrachloride (CCl_4), a fully chlorinated solvent. See Chlorofluorocarbon (CFC) solvents; Hydrochlorofluorocarbon (HCFC) solvents.

Chlorofluorocarbon (CFC) solvents (cleaning) Solvents containing chlorine and fluorine. Used in removing non-polar contaminants such as oils. Effective solvents but regulated because of health and environmental concerns. Examples: CFC-11 (CCl_3F); CFC-12 (CCl_2F_2); CFC-113 ($CF_2ClCFCl_2$). See Chlorinated solvents; Hydrochlorofluorocarbon (HCFC) solvents.

Chromate conversion Treatment of a metal surface with a hexavalent chromate solution to form a protective (corrosion-resistant) metal-chromate surface layer.

Chromium, decorative (electroplate) A thin chromium layer designed to improve appearance but not to impart corrosion or wear resistance. Often uses trivalent chromium baths.

Chromium, hard chrome (electroplate) A thick chromium layer designed to impart corrosion and wear resistance to the surface. Usually uses hexavalent chromium baths.

Chromium, hexavalent (electroplating) Chromium with a plus six charge. Very destructive to sewage disposal plants by killing bacteria.

Chromium, trivalent (electroplating) Chromium ions with a plus three charge.

Chromize The process of reacting a metallic surface with chromium to form a high-chromium alloy surface region.

Cladding The covering of a surface by a solid layer of a second material, then bonding the two together by temperature and pressure.

Cleanroom, materials for (cleaning) Materials that do not introduce particulates or vapors into the clean area. Examples: Non-linting cloth and paper; Stainless steel rather than vinyl furniture covering; Ink pens rather than carbon pencils.

Cleaning (cleaning) Reduction of the amount of contamination on a surface to an acceptable level.

Cleaning, alkaline (cleaning) A basic cleaner that cleans by saponifying oils. Alkaline cleaning is often followed by an acid rinse to neutralize the adhering alkaline material and remove non-soluble precipitates formed by reaction with the alkaline material.

Cleaning, external (cleaning) Cleaning done external to the deposition chamber.

Cleaning, gross (cleaning) Cleaning process designed to remove all types of surface contaminants, generally by removing some of the underlying surface material.

Cleaning, in situ (cleaning) Cleaning done in the deposition chamber.

Cleaning, plasma (cleaning) Cleaning done using an inert or reactive gas plasma either as an external cleaning process in a **Plasma cleaner** or as an in situ cleaning process in the deposition system. See Glow bar.

Cleaning, solvent (cleaning) Cleaning using a solvent that takes the contamination into solution. See Solubility test; Specific cleaning.

Cleaning, specific (cleaning) Cleaning process designed to remove a specific contaminant. Example: Removal of a hydrocarbon contaminant by oxidation.

Cleaning, sputter (cleaning) A gross, in situ cleaning process where the substrate surface is sputtered prior to the film deposition.

Cleaning, wipe-down (cleaning, vacuum technology) Cleaning by wiping with a wet lint-free, low-extractables pad containing a solvent such as alcohol. The wet surface picks up particulates and the solvent takes contamination into solution. Anhydrous alcohol is often used as a wipe-down fluid since it will displace water and will rapidly vaporize.

Cleaning procedure, RCA (semiconductor processing) A specific cleaning procedure designed to clean silicon wafers. A variation of the procedure is called the **Modified RCA cleaning procedure**.

Cleaving (cleaning) The process of introducing a fracture in a single-crystal material that follows a crystallographic plane. One method of producing a clean surface in vacuum.

Closed-loop system (cleaning) A cleaning line where the cleaners and rinsing agents are recycled so that there is very little dilute liquid waste generated. Contaminants are in the form of solids on filters or as concentrated liquid wastes. See Enclosed system.

Cluster tool (semiconductor processing)   An integrated processing system that is environmentally isolated such that with an appropriate transport mechanism the substrate can be moved from one processing chamber to another. Typically the transport can be used to position the substrate randomly in processing chambers clustered around a central chamber containing the transport mechanism. See In-line deposition system.

Coarse vacuum (vacuum technology)   Vacuum in the range of atmospheric to about 10^{-3} Torr. Also called **Rough vacuum** (preferred). See Rough vacuum.

Coat (garment, cleaning)   Outer clothing used to contain particulates generated on the body by presenting a barrier to air flow away from the body using a closely woven cloth or a solid fabric. Open at the bottom so particulates drop to the floor. See Bunny suit.

Coating   Term applied to overlaid material on a surface greater than 0.5 to several microns in thickness. Sometimes used synonymously with film. See Overlay; Thin film; Surface modification.

Coating, first surface   Coating on the side of the substrate on which the incident radiation impinges. Also called **Front surface coating**. Example: First surface mirror.

Coating, second surface   Coating on the side of a transparent substrate opposite the side on which the incident radiation impinges. Also called a **Back surface coating**. Example: Back surface mirror.

Coefficient of adhesion   The ratio of the force needed to pull surfaces apart to the force used to push them together.

Coefficient of friction (vacuum technology)   The ratio of the force parallel to the direction of motion needed to start movement (**Static friction**) or continue movement (**Dynamic friction**), to the load applied normal to the direction of motion. The higher the coefficient of friction, the more likely the galling and the generation of particulate contamination.

Coefficient of thermal expansion (CTE) (film formation)   The linear expansion (generally positive) as a function of increasing temperature.

Cohesion   The chemical bonding between like atoms in a bulk material.

Cohesive energy   The force (pounds/inch^2 or newtons/meter^2) needed to separate a bulk material and form two surfaces.

Coil source (evaporation)   A thermal evaporation source in the form of a coil, usually of stranded wire, that is wetted by and holds the molten evaporant material and allows deposition in all directions. See Evaporation source.

Coining (substrate)   Impressing a design into a surface by forcing a hardened die into the surface.

Cold cathode   A non-thermoelectron-emitting cathode that emits electrons, usually by secondary electron emission under ion bombardment or by radioactive decay. See Field emission.

Cold cathode ionization gauge (vacuum technology)   An ionization-type vacuum gauge where the electrons for ionization are usually produced by a secondary electron emitting surface or a radioactive material. Often uses a magnetic field to increase the path length of the electrons.

Cold cleaning (cleaning)   Cleaning performed at room temperature.

Cold light   Radiation from which the infrared flux has been substantially reduced.

Cold mirror (optics)   A thin film structure that reflects shorter wavelengths (typically visible) while transmitting longer wavelengths (infrared). See Heat mirror.

Cold trap (vacuum technology)   A baffle that operates by condensing vapors on a cold surface.

Cold welding   The bonding of metals at a low temperature, generally due to removal or disruption of the oxides on the metal surfaces. See Galling.

Collar, feedthrough (vacuum technology)   A short metal cylinder on which feedthroughs are mounted and located between the baseplate and the bell jar. Provides a sealing surface for both the baseplate and the bell jar.

Collimated sputter deposition (PVD technology)   Reduction of the non-normal flux from a sputtering target by using a honeycomb-shaped mechanical filter between the target and the substrate. Used to increase the throwing power in covering high-aspect-ratio surface features.

Colloid   Dispersion of small particles in a second material. In a fluid the particle suspension is controlled by collisional forces rather than gravity.

Color   The optical property (generally using reflected wavelengths) of a surface that stimulates color receptors in the human eye. The perception of color is sensitive to the illumination used and the individual observing the color. Color is quantified using the parameters L*, a*, and b*, where L* is the luster or brightness of the coating, a* is the color content from green to red (wavelength and amplitude), and b* is the color content from blue to yellow (wavelength and amplitude). See Brightness; Commission International de l'Eclairage (CIE).

Colorimetric imaging (characterization)   A method of locating pinholes in a film by reaction of the exposed substrate to a chemical to form a colored corrosion product that can be visually observed.

Columbium   The element niobium.

Columnar morphology (film formation)   The morphology that develops with thickness due to the development of surface roughness due to preferential film deposition on high points on the surface. The columnar morphology resembles stacked posts and the columns are not single grains. See Macrocolumnar morphology.

Comets   The visual trail in the deposition system left by molten globules emitted from a thermal vaporization or arc vaporization source. See Spits; Macros.

Commission International de l'Eclairage (CIE) (International Commission on Illumination)   The organization that provides standards of color for its measurement and specification.

Comparative test (characterization)   A test to compare a film property to a standard or to previous results without providing an absolute value. Comparative tests are often used in production to ensure product reproducibility.

Complex ion (electroplating)   An ion composed of two or more ions or radicals, each of which can exist independently.

Complexing (electroplating)   Attaching a metal ion to a larger ion so that its response to the electric field does not depend on the metal ion. Example: By complexing both lead and tin, a Pb-Sn solder alloy can be electrodeposited.

Complexing agents (electroplating)   Chemical agents, such as cyanides, that are used for complexing.

Compliant layer (adhesion)   An intermediate layer that can distribute the stress that is applied and prevent high stress-loads at the interface. The compliant layer can be of a porous material or an easily deformed material. See Buffer layer.

Composite material   A material composed of particles, precipitated grains, or fibers of one material dispersed in a matrix of another material. Example: Fiberglass and dispersion strengthened steel.

Compound, chemical   Material formed when two or more elements combine to form a phase with a specific crystalline structure and a specific composition (with the possibility of some variability in elemental ratios). Examples: SiO_2 and $SiO_{1.8}$ (silica and substoichiometric silica).

Compound-type interface (film formation)   When the interfacial material (interphase material) that has been formed during the deposition of A onto B, along with subsequent diffusion and reaction, consists of a compound of A and B such as an oxide. See Interface.

Compression ratio (vacuum pump)   The ratio of the outlet pressure to the inlet pressure of a vacuum pump at zero flow using a specified gas.

Compressive stress, film (film formation)   A stress resulting in the atoms being closer together than they would be in a non-stressed condition. Compressive stress tries to make the film material expand in the plane of the film. See Tensile stress.

Condensation energy (film formation)   The energy released upon condensing an atom or molecule from the vapor. See Heat of vaporization.

Conditioning, target (sputtering)   Removal of the surface contamination, such as oxides, and degassing of the target material before sputter deposition begins.

Conditioning, vacuum surface (vacuum technology)   The treatment of a vacuum surface to make the system more amenable to vacuum pumping. Treatment can include: plasma cleaning, sputter cleaning, heating, UV desorption, and/or hot-gas flushing.

Conductance (vacuum technology)   The measure of the ability of a part of a vacuum system to pass gases or vapors from the inlet to the outlet under steady state conditions. The units of conductance are Torr-liters/s of flow per Torr of pressure difference.

Conductance, parallel (vacuum technology)   When there are conductance paths (C_1, C_2, - -) that are in parallel. The total conductance (C_{total}) is the sum of the individual conductances i.e., $C_{total} = C_1 + C_2 + C_3 + - - - -$.

Conductance, series (vacuum technology)   When there are conductance paths (C_1, C_2, - -) that are in series. The total conductance (C_{total}) is given by $1/C_{total} = 1/C_1 + 1/C_2 + 1/C_3 + - - - -$.

Conductance, transit (vacuum technology)   The ability of a specific atom or molecule to pass from one end of a path to the other.

Conductive heat loss   Heat flow occurring between a hot region and a colder region of a material without mass movement.

Conductivity, water (cleaning)   The measure of the ionic conductivity of water using probes spaced one centimeter part. Expressed in **Megohms**. See Ultrapure water; De-ionized water.

Cone formation (sputtering)   Features that develop on a surface being sputtered that are due to having a low-sputtering-yield particle on or in (inclusion) the surface. The particle shields the underlying material from being sputtered. The angle of the sides of the cone depend on the angular dependence of the sputtering yield of the bulk materials with the specific bombarding ion.

Confined-vapor source (evaporation)   Evaporation source where the vapor is confined in a cavity and the substrate, such as a wire, is passed through the cavity.

Conflat™ (CF) flange (vacuum technology)   A demountable shear-sealing flange that uses opposing knife-edges to shear into a soft metal gasket.

Confocal microscope   An optical microscope where the reflected light from out-of-focus areas on a surface are prevented from reaching the optical detector by a small aperture in the optical path. Often used with horizontal scanning and precision vertical motion to give a series of "confocal slices" and allow a computer-generated 3-D image of the surface with high magnification, high contrast, high resolution, and an apparent high depth-of-field. Also called the **Confocal Laser Scanning Microscope**.

Conformal anode (electroplating)   An anode made to conform to the shape of the cathode to keep the anode-to-cathode spacing constant.

Conformal target (sputtering)   A sputtering target made conformal to the shape of the substrate in order to keep a constant spacing.

Contact angle (film formation, adhesion)   The angle of contact between a fluid drop and a solid surface as measured through the liquid. In some cases the contact angle with a fluid of known surface energy can be used to measure the surface energy of the solid (dyne test). In some cases, the **Advancing contact angle** or the **Receding contact angle** is measured.

Contaminant (cleaning)   A material that is contaminating the surface.

Contaminant, non-polar (cleaning)   Contaminants that are not polar materials. Example: Oils. See Polar contaminants.

Contaminant, polar (cleaning)   Contaminants that are polar materials. Example: Ionic salts. See Non-polar contaminants.

Contamination (PVD technology)   Materials in the vacuum system in a concentration high enough to interfere with the deposition process or to affect the film properties in an unacceptable manner.

Contamination (vacuum technology)   The materials in the vacuum system that affect the pumpdown time and the ultimate pressure of the system as well as the residual contamination in the system. See Base pressure.

Contamination, external environment-related (contamination control)   Contamination brought in from the external processing environment. Example: Particulate contamination from dust.

Contamination, process-related (contamination control)   Contamination from the deposition process. Examples: Outgassing of evaporation source; Volatilization of hydrocarbons from contaminated evaporation material.

Contamination, system-related (contamination control)   Contamination coming from the deposition system. Examples: Backstreaming from pump oils; Particulates from pinhole flaking in the system.

Contamination control (cleaning)   The control of contamination and recontamination of a surface by controlling the sources of contaminants. Example: Cleanrooms control the amount of particulate matter available for recontamination but do not control vapors that can recontaminate the cleaned surface.

Contract coater (surface engineering)   A manufacturing facility that will coat items for any individual requiring their services. See In-house coater.

Contractometer (electroplating)   Instrument for measuring stress in an electroplated coating.

Control samples   Samples retained after processing has been performed to allow comparison with material at a later stage or after being placed in service. See Shelf samples.

Conversion, natural (substrate)   The natural reaction of a material to form a surface layer. Example: Oxidation of aluminum or silicon after the original oxide has been removed. See Chemical conversion.

Converting (web coating)   The conversion of bulk metallized film (polymer web), paper, or board into a final product such as packaging, labels, decorative products, etc. Converting can involve laminating, sealing, slitting, printing, etc.

Convertor (web coating)   A manufacturer that utilizes metallized web material to fabricate a product. See Converting.

Coordination number (crystallography)   The number of nearest-neighbor atoms to a point in a lattice or on a surface.

Copolymer   A mixture of two different monomers to form a polymer material that is a mixture.

Copper-beryllium (alloy)   A copper-beryllium (Cu : 2% Be) alloy that is much harder than brass. Used in spring-type electrical contacts.

Copyright (U.S.)   The protection given to the author of a work to prevent others from reproducing the work without permission. Since March 1, 1989, all "tangible means of expression" (written words, photos, art, etc.) are automatically copyrighted. This means that permission needs to be obtained from the originator or copyright assignee for use of all or a significant part of the work. See Patent, utility.

Corona discharge   Electrical breakdown of the gas near a surface due to a high electric field that exceeds the dielectric strength of the gas. Usually seen at high-field points such as tips but can be found over planar electrically insulating surfaces that have been charged by an rf field. Example: St. Elmo's fire seen in nature under high electric field conditions.

Corona treatment (surface modification)   Treatment of polymer surfaces in a corona discharge in order to give the surface a higher surface energy and make it more wettable.

Correction run (optics)   A method of salvaging a coating that is out of spectral specifications by the addition or removal of layers.

Corrosion   Production of an undesirable compound or surface effect by reaction with the ambient environment.

Corrosion, chemical   Corrosion by purely chemical means.

Corrosion, electrochemical   Corrosion either driven by or enhanced by the presence of an electric field.

Corrosion, galvanic   Corrosion at the contact between two different materials in the presence of an electrolyte. Example: The pitting corrosion that results between a carbide inclusion and the matrix in welded stainless steel.

Corrosion, intergranular   Preferential corrosion at or along grain boundaries.

Corrosion, pitting   Corrosion that results in a pit on a surface.

Corrosion, stress   Corrosion due to or enhanced by intrinsic stress in the material. Example: Stress corrosion of aluminum by chloride ions.

Corrosion inhibitors   Molecular species that prevent corrosion by adsorbing on a clean surface and presenting a barrier to the corroding species. Also called **Rust inhibitors**.

Corrosive fluid (cleaning)   A fluid having a pH of less than 2.0 or greater than 12.5.

Corundum (abrasive)   An impure form of aluminum oxide (Al_2O_3). See Sapphire.

Cosine Law, Knudsen's   The intensity of flux from a point source impinging on a flat surface normal to the direction to the point of emission is proportional to the cosine of the angle subtended by the source at the plane surface and inversely proportional to the square of the distance ($\cos Y/r^2$).

Cost of ownership (COO)   Full cost of equipment including capital costs, financing costs, maintenance costs, utilities costs, operation costs, space costs, etc.

Counterflow rinse (cleaning)   **See Cascade rinse.**

Coupling agent (adhesion)   An agent that reacts with two materials, often through different mechanisms, and allows bonding of the materials together. See Glue layer.

Covalent bonding   The chemical bond that is formed between two atoms that each contributes one electron. If the electrons are shared unequally it is a **Covalent polar bond**. Also called **Electron pair bond**. See Chemical bond.

Cracking pattern   The portion of the spectra from a mass spectrometer due to the breaking up of complex molecules by electron bombardment. Also called **Fragmentation pattern**.

Craze   The network of fine hairline cracks in a surface or coating of a brittle material due to stresses generated during drying or curing.

Creep (contamination)   The movement of an adsorbate over a surface

Creep (deformation)   The long-term permanent deformation of a solid under mechanical stress.

Critical backing pressure (vacuum technology)   The foreline pressure above which a high-vacuum pump will not operate efficiently.

Critical cleaning   I know of no better definition than "Cleaning what already looks clean." Also called **Precision cleaning**.

Critical diameter (molecules)   The diameter of the smallest pore through which the atom or molecule can pass. Examples: He = 0.2 nm, O = 0.29 nm, H_2 = 0.24 nm, N_2 = 0.3 nm, O_2 = 0.3 nm, Ar = 0.29 nm, SF_6 = 0.67 nm, H_2O = 0.26 nm.

Critical point   The temperature (**Critical temperature**) and pressure (**Critical pressure**) at which a liquid and its vapor have the same density and other properties, thereby becoming indistinguishable.

Cross direction (web coating)   The direction orthogonal to the direction the web is moving. Also called **Transverse direction**. See Machine direction.

Cross-section   The physical area in which an interaction can take place. Examples: Cross-section for physical collision (sum of the radii of the particles); Cross-section for electron-atom ionization; Cross-section for charge-exchange collisions.

Crossed fields   Where the electric and magnetic fields have a vector component at an angle to one another. This situation produces a force on a charged particle moving in this region that is orthogonal to the plane of both fields. See Drift.

Crossover pressure (vacuum technology)   The chamber pressure at which the vacuum pumping system is switched from the rough pumping mode (**Roughing**) to the high-vacuum pumping mode.

Crosstalk (sputtering)   When material from one sputtering target is deposited on another target.

Crowding (vacuum technology)   When there is so much fixturing in the chamber that the conductance, particularly for water vapor, is reduced to the point that concentration gradients are established in the chamber.

Crown glass (optical)   A low-dispersion, relatively low-index optical glass used in the converging elements of lenses. See Flint glass.

Crucible, electrically conductive (evaporation)   A crucible of an electrically conductive material such as carbon or TiB_2 plus BN that can be heated resistively or by accelerated electrons.

Crucible, evaporation   A container for holding molten material. See Skull.

Crucible, water-cooled (evaporation)   A crucible that is water-cooled and where the evaporant material is heated directly by an electron beam. See Hearth.

Cryocondensation (vacuum technology)   Adsorption on a cold surface which may or may not be covered with an absorbate material.

Cryogenic fluid   Fluid with a boiling point below -150°C.

Cryopanel (vacuum technology)   A vapor pump that operates by cryocondensation of vapors on a large-geometrical-area cold surface at a temperature between < -150°C where the vapor pressure of water is very low. Also called a **Meissner Trap**. See Cryopump.

Cryopump (vacuum technology)   A capture-type pump that operates by condensation and/or adsorption on cold surfaces. Typically there are several stages of cold surfaces. Typically one of the stages will have a temperature below 120 K. See Vacuum pump.

Cryosorption pump (vacuum technology)   A vacuum pump that operates by cryocondensation of gases on large-adsorption-area cryogenically cooled (< -150°C) surfaces. Also called a **Sorption pump**. See Vacuum pump.

Cryotrapping (vacuum technology)   The physical trapping of a gas in a porous material such as a zeolite or activated carbon when the surface mobility is low because of a low temperature.

Crystal structure (material)   The ordered arrangement of atoms in a solid material that is characterized by the spacing between atoms and the direction from one atom to another. The crystalline structure is comprised of repeating groups of atoms called **Unit cells**. Also called **Lattice structure**.

Crystal structure, body centered cubic (bcc)   A crystal structure where the basic building block is a cubic unit cell having atoms at each corner and one in the center of the cell.

Crystal structure, diamond   A crystal structure where each atom is at the center of a tetrahedron formed by its nearest neighbors. Example: Diamond.

Crystal structure, face centered cubic (fcc)   A crystal structure where the basic building block is a cubic unit cell having atoms at each corner and one in the center of each face.

Crystal structure, hexagonal close packed (hcp)   A crystal structure where in alternate layers of atoms the atoms in one layer lie at the vertices of a series of equilateral triangles in the atomic plane, and the atoms in the layer lie directly above the center of the triangles in the atomic plane of the next layer. Example: Beryllium.

Crystal structure, tetragonal   A crystal structure where the axes of the unit cell are perpendicular to each other and two of the axes are of equal length but the third is not of the same length.

Crystalline (material)   A material that has a defined crystal structure where the atoms are in specific positions and are specific distances from each other.

Crystallographic plane   One of many planes in a crystal structure that contains atoms. The areal density of the atoms and spacing between the atoms on the plane vary with direction. Also called **Atomic planes**. See d-spacing.

Curie   An amount of a radioactive nuclide that has 3.7 x 10^{10} disintegrations per second.

Curie temperature (T_c)   Temperature above which a ferromagnetic material loses its ferromagnetism. Examples: 627 K for Ni; 1043 K for Fe.

Curing, polymer   The conversion of a fluid containing monomers to a solid by polymerization. Curing may occur by reaction in a two-part system (Example: A two-part epoxy), thermal curing, electron-beam curing, ultraviolet radiation curing, etc. The curing operation can leave significant amounts of low-molecular-weight material in the solid material. See Undercuring.

Curling, film (adhesion)   When a film separates from the substrate and curls up due to non-isotropic stress through the thickness of the film.

Current density   Current per unit area. Example: 1 mA/cm^2 of singly charged ions equals 1.6 x 10^{16} ions per second per square centimeter. See Ampere.

Cut-off wavelength (optics)   The maximum wavelength at which a material or a filter will pass radiation.

Cut-on wavelength (optics)   The minimum wavelength at which a material or a filter will pass radiation.

Cyanide compound (safety)   Any of a group of toxic compounds containing the CN group, usually derived from the compound HCN.

Cyanoacrylate glue (vacuum technology)   A class of adhesives used to bond rubber materials. Example: Used to splice rubber O-rings. Also called **Super Glue.**

Cycle time, processing   The time for one complete processing sequence, including loading and unloading.

Cyclotron frequency (plasma)   Resonant adsorption of energy from an alternating electric field by electrons confined in a uniform magnetic field when the frequency of the electric field matches the oscillation frequency of the electrons in the magnetic field.

Cylindrical (hollow) magnetron (sputtering)   A hollow cylindrical tube, often with ends flared toward the interior, where a magnetic field confines the secondary electrons emitted from the inside surface to paths parallel to the axis of the tube (magnetron configuration). The flares prevent the loss of the electrons from the ends of the tube. See Magnetron.

d-spacing (crystallography)   The spacing between like atomic planes in a crystal lattice.

Dalton's Law of Partial Pressures (vacuum technology)   Dalton's Law of Partial Pressures states that the sum of all the partial pressures of gases and vapors in a system equals the total pressure. See Partial pressure.

Damage threshold (bombardment)   The energy at which radiation or bombarding particles will introduce damage to the atomic structure of a material, thus changing its properties.

Damascene pattern (semiconductor metallization)   Inlay of one material into another to provide a patterned flat surface. Structure is obtained in semiconductor processing when a material is deposited in vias and trenches on a surface, then the high areas are polished back to the original surface. See Chemomechanical polishing (CMP).

Dangling bonds   An unsatisfied chemical bond that is available to react with atoms or molecules. See Sensitization, surface.

Dark current (*Archaic*)   The current through a glow discharge tube when a portion of the discharge looks "dark."

Dark space, cathode (plasma)   The darker region of a plasma near the cathode surface where most of the potential drop in a DC diode discharge occurs. Region where electrons are being accelerated away from the cathode. Also called the **Cathode sheath**.

Dark space shield (plasma)   A grounded surface that is placed at less than a dark space width from the cathode in order to prevent establishing a discharge in the region between the two surfaces. Also called the **Ground shield**. See Paschen curve.

DC glow discharge (plasma)   The plasma discharge established between two electrodes in a low-pressure gas and in which most of the potential drop is near the cathode surface and a plasma region (positive glow) where there is little potential drop that can extend for an appreciable distance.

De-excitation (plasma)   The return of an electron in an excited state to a lower energy level, accompanied by the release of optical radiation. Also called **Relaxation**.

De-wetting growth (film formation)   When the nuclei tend to grow normal to the surface rather than laterally over the surface. See Wetting growth.

Deadhesion   The loss of adhesion. See Adhesion.

Debug   To eliminate the initial problems in an electronic circuit or a software program.

Deburring   The removal of burrs that are formed during deformation or cutting operations.

Decarburizing   The loss of carbon from a carbon-con-

taining compound or alloy. The loss may be due to diffusion, vaporization, or chemical reaction.

Decomposition reaction (CVD)   Deposition by decomposition of a chemical vapor precursor species. Example: Si from SiH_4.

Decorative coating   A coating whose function is to be decorative so that the properties of the coating of interest are primarily reflectivity, color, color distribution and texture. Often protected using a **Topcoat**.

Decorative/functional coating   A coating that has the requirements of a decorative coating but also improves some functional property such as abrasion, wear, corrosion, etc. Example: Decorative coating on a plumbing fixture or door hardware. See Functional coating; Decorative coating.

Deep ultraviolet (DUV)   Short-wavelength ultraviolet radiation.

Defects, film (film formation)   Any irregular feature of the film crystallinity, microstructure, or morphology that can affect the film properties. Examples: Pinholes; Voids; Column boundaries.

Defects, lattice (crystallography)   Any departure from crystalline order such as **Vacancies**, **Substitutional atoms**, **Interstitial atoms**, **Dislocations**, **Grain boundaries**, etc.

Defects, surface (substrate)   Any feature on the surface that disrupts the regularity and that might influence film growth, film properties, or film adhesion. Examples: Scratches; Microcracks; Electronic charge sites.

Deflected electron beam (evaporation)   An e-beam evaporation source where the electron beam is deflected out of the line-of-sight of the electron emitter to impinge on the surface of the charge. The e-beam can be focused and rastered over the surface of the charge during heating.

Deflocculants (cleaning)   Chemicals that are added to solutions to help maintain the dispersion of contaminants suspended in the cleaning medium.

Degas (fluids)   Removal of gases and vapors from a liquid, usually by heating or reduction in pressure above the fluid. Also called **Exosolution**. See Outgas.

Degassing rate   The rate at which gases or vapors leave a surface. Measured in Torr-liters/sec-cm^2 or grams/sec-cm^2. See Outgassing rate.

Degreaser, vapor (cleaning)   A cleaning system where the surface to be cleaned is placed in the hot vapor of the cleaning solvent. The vapor condenses on the surface, dissolving the contaminant, and flows off into the sump. When the part reaches the temperature of the vapor, condensation stops and the part is removed. In the old-style degreaser, which was open to the atmosphere, there was a spray wand that allowed spraying the part while in the vapor. See Degreaser, vapor, low-emission; Drying, vapor.

Degreaser, vapor, low-emission (cleaning) A degreaser where the cleaning solvent is contained in an enclosed cleaning chamber, then pumped away before the cleaning chamber is opened. The vapors are condensed and returned to the cleaning liquid sump.

De-ionized (DI) water (cleaning) Water in which most of the ions, which have a potential for reaction with cleaning materials and/or leaving a residue, have been removed. Often used (erroneously) synonymously with ultrapure water. For de-ionized water the electrical conductivity can be as low as 18.2 megohm-cm at room temperature. See Conductivity; Ultrapure water.

Deliquescent (vacuum technology) Material that reversibly absorbs and desorbs water from the air. Example: NaCl (common table salt).

Demister (vacuum technology) A baffle on the exhaust of an oil-sealed mechanical pump used to condense oil vapors to reduce the loss of oil from the pump.

Denatured alcohol (cleaning) Ethyl (grain) alcohol that has be rendered unfit to drink by the addition of another material (**Denaturant**).

Density The mass per unit volume (g/cm³) of a material. See Specific gravity.

Density gradient column (characterization) A liquid column in which the density of the liquid is varied by having a temperature gradient. An object immersed in the liquid will float at a level where its density matches that of the fluid.

Deposition rate Mass or thickness of material deposited per unit time. Measured in micrograms per cm² per second, nanometers per second, or Ångstroms per second.

Deposition system (PVD technology) A vacuum system used for physical vapor deposition processing.

Deposition system, cluster-tool (semiconductor processing) A load-lock vacuum system that has random access to several processing modules from the loading/transfer chamber.

Deposition system, batch See Deposition system, direct-load.

Deposition system, direct-load A system where the processing chamber is opened to the ambient each time the fixture is placed into or removed from the chamber. Also called a **Batch system**.

Deposition system, dual-chamber A chamber that has two separate sections separated by a low conductance path. The sections may be independently pumped or there may be two different gas pressures in the sections. This allows high gas load operations, such as unrolling a web, to be performed in a section separate from the film deposition section.

Deposition system, in-line A series of sequential vacuum modules beginning and ending with load-lock chambers that allows the substrate to enter one end and exit the other end without reversing direction.

Deposition system, load-lock A system that has a chamber intermediate between the ambient and the deposition chamber that allows the substrate to be outgassed, heated, etc., before being placed in the deposition chamber. The substrates are passed from the load-lock chamber into the deposition chamber through an **Isolation valve** using **Transfer-tooling**.

Deposition system, web coater Specialized direct-load deposition system used to coat web material that is often on very large, heavy rolls. Often a dual-chamber system. Also called a **Roll coater**.

Depth profiling (characterization) The determination of the elemental composition as a function of distance from the surface. The analysis may be destructive (e.g., sputter profiling using Auger Electron Spectroscopy [AES]) or non-destructive (e.g., profiling using Rutherford Backscattering Spectrometry [RBS]).

Descale (cleaning) The chemical or electrochemical removal of thick oxide layers (scale) from a surface.

Desiccant (cleaning) A chemical that has a great affinity for water and will reduce the relative humidity in its surroundings to a very low value.

Design rule (general) A design requirement or the relationship that has been established between design parameters and a desired result. Example (sputter deposition): For planar magnetron sputter deposition, the length of the sputtering "racetrack" should be the length of the fixture holding the substrates in front of the sputtering target plus twice the target-substrate distance in order to get reproducible film properties from the top, middle, and bottom positions on the fixture.

Design rule (semiconductor processing) Spacing between interconnect metallization lines (e.g., 0.35 micron design rule).

Desize (cleaning) Removing the **Sizing** (lubricant) from a cloth by washing in hot water.

Desorption To remove gases and vapors from the surface of a material, usually by heating but also by electron impact, ion impact, etc. See Outgas.

Desorption energy The amount of energy necessary to cause an atom or molecule to vaporize from the surface of a material. See Thermal desorption spectroscopy.

Detergent (cleaning) A substance that reduces the surface tension of water, concentrates at the water-oil interface, and takes oils into suspension (emulsifies them). Detergents can be of several types: **Anionic detergents, Cationic detergents**, or **Non-ionic detergents**.

Detonation gun deposition (thermal spray) A thermal spray process in which the particles are melted in an explosion front and propelled to a high velocity in a "gun barrel."

Devitrification Crystallization of a glassy material.

Dew point, water The temperature at which the vapor pressure of water reaches saturation and the vapor begins to condense into a fluid. See Humidity.

Dewar vessel (vacuum technology) A vacuum-insulated container commonly used to contain liquefied gases.

Diameter, atomic and ionic The physical diameter of atoms and ions. Examples: $O_2 = 2.98$Å, $O^o = 0.60$Å, $O^{-2} = 1.32$Å; $Sn^o = 2.8$Å, $Sn^{+4} = 1.42$Å, $Sn^{-4} = 5.88$Å.

Diamond (abrasive) The crystalline form of carbon that is very hard. Commonly available in abrasive particle sizes down to 0.25 micron.

Diamond-like carbon (DLC) An amorphous carbon material with mostly sp^3 bonding that exhibits many of the desirable properties of diamond but does not have the crystal structure of diamond.

Diamond point turning (substrate) Machining a metal using a light-cut with a very sharp, wear-resistant point on a diamond tool, thus obtaining a very smooth, mirror-like as-machined surface.

Diaphragm pump A gas or fluid pump that operates by the periodic expansion and reduction of a chamber volume by the action of a piston-actuated flexible (usually polymeric) diaphragm. In vacuum applications the diaphragm pump can be used at pressures down to 10 Torr at the inlet with an exhaust to atmospheric pressure.

Diatomaceous earth (cleaning) Soft material (88% silica, balance calcium carbonate) composed of the skeletons of small prehistoric aquatic plants. Used as a mild abrasive and as a filtration material. When the calcium carbonate is removed by acid washing, the material is used as a fine silica abrasive.

Dichroic coating An optical coating that reflects certain wavelengths and allows others to pass through. Examples: Heat mirror; Sunglass coatings. See Ophthalmic coatings; Band-pass coatings.

Die (semiconductor) The conductor circuit pattern on the surface of a chip which is connected to a printed circuit board or chip carrier by wires (to a **Lead-frame**) or solder bumps (**Flip-chip bonding**).

Dielectric An electrically insulating material that has a dielectric constant greater than one.

Dielectric constant (material) The ratio of the capacitance of a capacitor constructed using the dielectric material as the insulator between the electrodes, to a capacitor using vacuum between the two electrodes.

Dielectric material A material that is an electrical insulator. A material that has little optical adsorption.

Dielectric strength The voltage gradient that can be tolerated by a material without an electrical breakdown (arc) through the material.

Differentially pumped (vacuum technology) A system or component in which one region is pumped differently from another. This may be done using different pumps or by different pumping manifolds. Example: Differentially pumped, dual O-ring sealed, mechanical motion feedthrough where the space between O-rings on the shaft is pumped.

Diffuse reflection Optical reflection in many directions. Diffuse reflection is due to surface roughness on the order of the wavelength of the light or greater. Also called **Non-spectral reflection**. See Scatterometry; Spectral reflection.

Diffusion The movement of one atomic, ionic, or molecular species through another due to a concentration gradient or an electric field gradient.

Diffusion-type interface (film formation) When the interfacial material (interphase material), which has been formed during the deposition of A onto B along with subsequent diffusion, consists of an alloy of A and B with a gradation in composition. See Interface; Kirkendall porosity; Interphase material.

Diffusion pump (DP) (vacuum technology) A compression-type vacuum pump that operates by the collision of heavy vapor molecules with the gas molecules to be pumped, giving the gas molecules a preferential velocity toward the high-pressure stages of the pump. See Vacuum pump.

Diffusivity The rate of diffusion across an area. Also called the **Diffusion coefficient**.

Diluent gas (CVD) A gas that does not enter into the deposition process but is used to control the partial pressure of the precursor gas at a given total gas pressure. Also called **Carrier gas**.

Diluent gas (vacuum technology) Dry gas used to dilute a vapor-containing gas to the point that the vapor will not condense during compression in a mechanical pump. See Ballast valve.

Dimers A vapor species consisting of two molecules.

Dioctyl phthalate (DOP) (contamination control) A chemical used to generate the white fog that is used to test HEPA filters.

Dip coating Where the part is dipped into a fluid and the fluid is allowed to drain off the part. The viscosity of the fluid determines the coating thickness. See Flow coating.

Direct current (DC) A voltage waveform where the polarity is the same or zero at all times. See Pulsed DC.

Disappearing anode effect (sputtering) In reactive deposition of electrically insulating films, the surfaces in the deposition chamber become covered with an insulating film and the electron flow to the grounded surface (anode) must change position as the surfaces become coated.

Discharge pressure (vacuum technology) The pressure at the outlet of the high-pressure stage of a vacuum pump. Also called **Exhaust pressure**. See Foreline pressure.

Dislocation, lattice (crystallography) A line of displacement of atoms in a lattice. Often formed during mechanical stress to relieve some of the stress.

Dispersion (cleaning) Breaking up big particles into small particles that can be suspended in water. Alkaline silicates and alkaline phosphates are used as dispersion agents in some cleaning formulations.

Dispersion (optical) The sensitivity of the optical properties, such as index of refraction, of a transparent optical material to the wavelength of the radiation being transmitted.

Dispersion strengthening When a small amount of a second phase in the form of small particles is dispersed in a matrix and strengthens the material. The particles may be mixed with the material in the melt or be formed by reaction and precipitation after the solid has been formed.

Displacement plating When an ion in solution that has a less negative electrochemical potential than the atom of the solid spontaneously displaces the atom of the solid and deposits it on the solid. Examples: Au (+1.50 volts) plating onto Cu (+0.52 volts); Pb (-0.126 volts); or Sn (-0.136 volts) (from solder) plating on Al (-1.67 volts). Also called **Immersion plating**. See Electrochemical series.

Disproportionation reactions (CVD) A reaction where the oxidation state of the element both increases and decreases through the process. Process can be use to purify materials.

Dissociation (plasma chemistry) Separation of a molecule into two or more fragments due to collision (Example: collision of a electron with a molecule) or the adsorption of energy (Example: photodissociation). See Fragmentation.

Dissociative attachment (ionization, plasma chemistry) When a molecule combines with an electron, loses a fragment, and becomes a negative ion. Example: $SF_6 +$ $e^- => SF_5^- + F$ (SF_6 is a good electron scavenger in a plasma.)

Documentation (manufacturing) The documentation that is maintained in order to know what was done during the processing and the status of the processing equipment. This enables reproducible processing to be performed and assists in failure analysis at a later date.

Documentation, log A dated document detailing who, when, and what was done (i.e., **Log, calibration**; **Log,** maintenance; **Log, run time**).

Documentation, manufacturing processing instruction (MPI) Detailed instructions for the performance of each operation and the use of specific equipment, based on the specification, that apply to each stage of the process flow. MPIs are developed based on the specifications.

Documentation, process flow diagram (PFD) A diagram showing each successive stage in the processing sequence including storage, handling and inspection. A PFD is useful in determining that there are MPIs that cover all stages of the processing.

Documentation, specifications (Specs) The formal document that contains the "recipe" for a process and defines the materials to be used, how the process is to be performed, the parameter windows and other important information related to safety, etc. Information on all critical aspects on the process flow sheet should be covered by specifications.

Documentation, travelers Archival document that accompanies each batch of substrates detailing when the batch was processed and the specifications and MPIs used for processing. The traveler also includes the **Process sheet**, which details the process parameters of the deposition run. Also called a **Run-card** in semiconductor processing.

Dog-boning (electroplating) When the deposit builds up at a faster rate at high field regions, such as at corners, compared to a flat region.

Donor, electrical An impurity (dopant) that increases the number of free electrons in the material. See Acceptor.

Dopant (glass) A chemical element that is added to give color to a glass.

Dopant (semiconductor) A chemical element added in small amounts to a semiconductor material to establish its conductivity type and resistivity. Examples: Phosphorus; Arsenic; Boron. See Donor; Acceptor.

Dose (ion bombardment) The total number of bombarding particles per unit area.

Double bond A type of chemical bonding where two pairs of electrons are shared equally between two atoms. Symbolized by (=). Example: C=O.

Down-time The amount of time that a pump or system is not operational due to failure or maintenance requirements. See Up-time.

Downstream region (plasma technology) See Afterglow region.

Drag finishing (substrate) Polishing a surface by pulling individual parts through an abrasive media. This prevents part-to-part contact that can cause damage.

Drag pump, molecular (vacuum technology) A vacuum pump that imparts a preferential motion to a gas molecule by the friction between the gas and a high-velocity surface. See Vacuum pump.

Drag-out (cleaning, electroplating)   The transfer of fluid from one tank to the next by virtue of the liquid material retained on the surface. Drag-out often necessitates a rinse step between the two tanks to prevent contamination of the second tank.

Drift, EXB   The motion of an electron in a direction normal to the plane defined by the electric and magnetic field vectors.

Drift, gauge (vacuum technology)   The change of calibration of a sensor with time or use.

Dry (cleaning)   Removal of water from a surface after processing, hopefully without leaving a residue. See Water spot.

Dry, blow-off   Removal of water by blowing it off a surface with a high-velocity gas stream. See Air knife.

Dry, displacement   Removal of water by taking it into solution with another fluid (**Drying agent**), such as anhydrous alcohol, that has a rapid drying rate when pure.

Dry, hot gas   Using a hot dry gas to dry a surface.

Dry, spin   Drying by spinning the surface at a high velocity and slinging the water off the edges. See Spin coat.

Dry, vacuum   Drying under vacuum to aid in the removal of moisture.

Dry, vapor (cleaning)   A cleaning system where the surface to be dried is placed in the hot vapor of the drying agent. The vapor condenses on the surface dissolving and displacing the water and flowing off into the sump. When the part reaches the temperature of the vapor, condensation stops and the hot part is removed where it dries rapidly. See Drying, vapor, low-emission; Degreasing, vapor.

Dry gas   A gas with a very low dew point for water vapor. Examples: Dry air with a dew point of -100°C; dry hydrogen with a dew point of -70°C (commercial grade dry hydrogen).

Dry process   A process that uses no fluids. Often desirable in context of waste disposal.

Dry pump (vacuum technology)   Vacuum pump that uses no (or little) oil, which can become a source of contamination. See Vacuum pump.

Dry pumping (contamination control)   Vacuum pumping using one or more dry pumps to avoid the possibility of oil contamination. Example: A turbopump with a molecular drag stage backed by a diaphragm pump.

Drying, vapor, low-emission (cleaning)   A drying system where the drying agent is contained in the drying chamber, then pumped away before the drying chamber is opened. The vapors are condensed and returned to the drying agent **Sump**.

Drying agent   A fluid used to displace fluids that have potential residue materials and that will vaporize quickly from the surface. Examples: Anhydrous alcohol; Perfluoro-N-methyl morpholine plus 0.2% surfactant (3M PF-5052 DS "spot-free" drying agent).

Dual-containment piping   A configuration where an exterior pipe surrounds the supply pipe that carries a high-purity or hazardous gas or liquid. The outer volume can be evacuated and monitored for safety.

Ductile fracture   Fracture that is accompanied by appreciable plastic deformation. See brittle fracture.

Ductile material   A material that undergoes appreciable plastic deformation before failure. See Brittle material.

Ductility   The ability of a material to plastically deform under applied stress. See Elongation; Elasticity.

Dummying (electroplating)   Removing tramp elements from the electrolyte by plating them out before the product is coated.

Duplex steel (substrate)   A simple alloy of iron and carbon, perhaps with a little Si, Ni, or Mn. The alloy has high ductility and easy formability, and is used in stamping parts such as auto fenders. Also called **Dual-phase steel**.

Durometer   An instrument for measuring the elastic deformation of a material (elastomer) under a controlled load.

Dust balls (cleaning)   Balls of lint that accumulate lint by rolling around on the floor in air currents. Also called **Dust bunnies**.

Dusters (cleaning)   Soft mop-like dusters, often made of electret material, used to collect dust and not generate particulates.

Duty cycle (equipment)   The ratio of the working time to the total time of a piece of equipment.

Duty cycle (process)   The ratio of the time of the processing sequence, such as pumpdown to a base pressure in the total process cycle time or actual sputtering time per voltage cycle in pulse power sputtering, to the process cycle time of interest.

Dwell (cleaning)   The time the part remains in a specific cleaning stage. Example: In the vapor of a vapor degreaser. See Soak.

Dyne test (surface)   Determining the surface energy of a polymer by applying fluids with known surface energies to the surface and monitoring the contact angle or by marking with materials (e.g., **Dyne-test marker pens**) having progressive (30-60 dyne/cm) surface energies.

E-beam evaporation (PVD technology)   Evaporation in a good vacuum using a focused high-energy, low-current electron beam as the means of directly heating the material to be evaporated.

E-beam melting (metallurgy)   Melting an alloy in a good vacuum using a high-voltage electron beam. See Vacuum Arc Melting; Vacuum Induction Melting.

E-coat   Coating applied by electrophoretic deposition (electrocoating). Usually an organic material but may be particles of inorganic materials such as glass. See Electrocoating.

E-diagnostics   Remote monitoring of system or process performance.

E-inventory   Remote monitoring of inventory.

Eddy current (thickness measurement)   An AC magnetic field from a probe induces magnetic eddy currents in an electrically conductive material (film or substrate) that generates an opposing magnetic field, which alters the circuit reactance and output voltage of the probe. Technique can be used to measure non-magnetic films on ferromagnetic material and non-conductive coatings on non-ferrous materials.

Edge filter (optics)   A filter that transmits at wavelengths longer than its **Cut-on wavelength** or shorter than its **Cut-off wavelength**.

Effusion cell   A thermal vaporization source that emits vapor through an orifice from a cavity where the vapor pressure is carefully controlled by controlling the temperature. Example: Used in MBE processing. Also called a **Knudsen cell**.

Elastomer   Material that is elastic or rubber-like, i.e., under stress it can deform to a large extent, exert a restoring force, then return to its original shape when the deforming force is removed.

Elastomer seal (vacuum technology)   A deformation seal that is made from an elastomer such as Viton™, Butyl-rubber, or silicone rubber. See Seal, spring-loaded.

Electret (cleaning)   A polymer material that has a permanent electric polarization charge. Usually formed by deformation of a polymer in an electric field.

Electrical evaporation (*Archaic*)   An old (< 1910) term for sputtering.

Electrical resistance   The electrical resistance (R) of a conductor is given by: $R = \rho\, L/A$ where ρ is the bulk resistivity in ohm-cm, L is the length of the conductor in cm, and A is the cross-sectional area of the conductor in cm^2. See Sheet resistivity.

Electrochemical polishing   Smoothing a surface by a combination of chemical polishing (selective chemical dissolution of high points) and electropolishing (selective off-plating of high points).

Electrochemical series   The relationship of materials as to their electrode potential (tendency to lose electrons as related to a platinum/hydrogen electrode, i.e., electrode potential). Also called the **Electromotive Series**.

Electrochromic film (optics)   A thin film structure that changes optical density under the influence of an applied electric field.

Electrocleaning (cleaning)   Removal of a material from substrate that is made the electrode (cathode or anode) of an electrolysis cell.

Electrocoating   The deposition of charged particles (paint, glass, etc.) from an electrolyte under an applied voltage. The deposition can either be on the cathode (**Cathodic electrocoating**) or the anode (**Anodic electrocoating**). Also called **Electrophoretic deposition**.

Electrode   An electrically conductive surface that is active in carrying an electric current. See Cathode; Anode.

Electrode potential   The voltage generated when a material is immersed in an electrolyte and usually referred to a standard platinum/hydrogen electrode used as the zero potential. See Electrochemical series.

Electrodeposition   The deposition of ions from a solution on the cathode of an electrolysis cell. Generally the ions lost from the solution are replenished by dissolution of the anode. Also called **Electroplating**.

Electroetching (cleaning)   Electrolytic removal of material from an anodic surface without the presence of a passivating surface layer. See also Electropolish.

Electroforming (electroplating)   The generation of a free-standing structure by electrodeposition on a shaped mandrel, then removing the mandrel. See Vapor forming.

Electrographic printing (characterization)   A method of locating pinholes in a film by reacting the exposed substrate with a wet chemical in an applied electric field to form a colored corrosion product that can be visually observed.

Electrography   Forming an image by the attraction of electrically charged "toner" to a selenium (or other photosensitive material)-coated drum that has been charged by exposure to an optical image, transferring the toner to paper, then fusing the toner to the paper with heat. Also called **Xerography**; **Electrophotography**.

Electroless plating   Deposition of a coating from a solution by use of a reducing agent in the solution rather than an external-applied electrical potential. Also called **Autocatalytic deposition, Autodeposition, Autophoretic deposition**. Example: Electroless Ni, Cu.

Electrolysis   A method by which chemical reactions are carried out by passing an electrical current through an electrolyte. Example: Electrolysis of water to form hydrogen and oxygen.

Electrolyte   A solution or gel containing a chemical compound that will conduct electricity by virtue of dissociation of the chemical compound into ions that are mobile in the media.

Electrolytic anodization (surface modification) Oxidation of the surface of a material at the anode of an electrolysis cell. See Anodization.

Electrolytic conversion The production of a compound layer on the surface of an electrode in an electrolysis cell. Example: Anodization.

Electromagnetic interference (EMI) shielding Thick deposits of metal to prevent electromagnetic radiation from penetrating into or out of a container and affecting electronic components.

Electromigration (semiconductor) The movement of atoms in a metallic conductor stripe under high current conditions ($> 10^6$ A/cm^2 in aluminum).

Electromotive series See Electrochemical series.

Electron Elementary particle having a negative charge and a mass of approximately $1/1840$ that of a hydrogen atom.

Electron beam (e-beam) (evaporation) Heating and evaporation of a material by an electron beam. The electron beam generally has a low-current of high-energy electrons, is directed to the surface of the material to be evaporated, and may be rastered over the surface during heating. Electron beams of low-energy and high current can be used to evaporate material, but the term "e-beam" is generally applied to a beam using high-energy electrons.

Electron cyclotron resonance (ECR) plasma source (plasma technology) A plasma source where the microwave energy, which has a resonant frequency of the electron in a magnetic field, is injected into the plasma-generating region through a dielectric window. See Plasma source.

Electron impact excitation (plasma chemistry) Excitation of an atom or molecule by electron impact. See Excitation.

Electron impact fragmentation (plasma chemistry) Fragmentation of a molecule by electron impact.

Electron impact ionization (plasma chemistry) Ionization of an atom or molecule by the impact of an electron, causing the loss of an electron. See Ionization.

Electron spectroscopy for chemical analysis (ESCA) (characterization) A surface analytical technique where the probing species are X-rays and the detected species are photoelectrons. The technique allows identification of species on the surface and the chemical binding energy. Also called **X-ray photoelectron spectroscopy (XPS).**

Electron temperature (plasma) A measure of the average kinetic energy of electrons in a plasma.

Electron volt (eV) The amount of kinetic energy imparted to a singly charged particle when accelerated through a potential of one volt. Equal to 1.602×10^{-19} Joules. A particle with 1 eV of energy has a temperature equivalent to about 11,600 K.

Elecronegativity The relative propensity for an atom to lose or gain an electron as given by the **Electromotive series.**

Electronic filter (cleaning) An air filter that ionizes particulates in a high electric field and the charged particles are then attracted to electrically grounded surfaces. See Electrostatic filter; Mechanical filter.

Electronic grade material A purity grade for materials that are to be used in electron devices such as electron tubes.

Electro-optical property (film) A property of a film, such as optical transmission or color, that is affected by electric fields.

Electrophoresis The migration of large electrically charged solid particles or liquid droplets (emulsion) in a fluid medium under the influence of an electric field. Also called **Cataphoresis.**

Electrophoretic deposition Deposition of larger-than-ion charged particles from a solution by electrophoresis. Particles can be of glass, polymer, liquid, etc. Deposition is generally on the cathode but may be on the anode of the electrolytic cell.

Electroplating Deposition of ions of a material from an electrolyte on the cathode of an electrolysis cell. Generally the ions being removed are replenished by dissolution of an anode of the material being deposited. Also called **Electrodeposition.**

Electropolishing Electrolytic removal of material from the high points on an anodic surface with concurrent passivation (usually by phosphates) of the smoothed areas. See Electroetching.

Electrostatic charge The potential on an electrically isolated part or surface.

Electrostatic filter (cleaning) A filter that attracts charged particles by virtue of a permanent electrostatic charge on the filter material. See Electret; Electronic filter; Mechanical filter.

Electrostatic spraying Coating using a spray of liquid or solid particles having an electric charge so that they can be directed to the substrate by an electric field.

Ellipsometry The technique for determining the optical constants or thickness of a film by determining the change in phase and amplitude of the electrical field vector of light reflected from the surface.

Embrittlement (metallurgy) The reduction in fracture toughness of a material by the addition of impurity atoms such as hydrogen or helium in high-strength steel, or mercury or indium in aluminum.

Emery   A natural abrasive material consisting of 55-75% aluminum oxide (Al_2O_3) and the rest being iron oxide and other impurities.

Emission spectrum, optical (plasma)   The de-excitation spectrum (color) of atoms and molecules in a plasma. The intensity of the peaks in the spectrum will change with changes in the plasma parameters.

Emulsification (cleaning)   To establish a stable suspension of particles in a fluid by coating them with a surfactant that prevents them from combining into large masses.   See Floculation.

Emulsion cleaner (cleaning)   A cleaning solution consisting of an organic solvent emulsion suspended in a water base.

Enabling technology (manufacturing)   Euphemism for the processes and equipment that work.

Enameling   A fusion coating where the coating consists of a glassy matrix, often containing a pigment, that bonds to the substrate surface. See Fusion coating.

Enclosed system (cleaning)   Cleaning, rinsing, and drying systems where the liquids are contained and the vapors are condensed and recycled. This reduces pollution generation. Examples: Vapor cleaners; Spray cleaners; Vapor dryers. See Closed-loop system.

End-Hall plasma source (plasma technology)   A plasma source that uses a thermoelectron emitter and a magnetic field to confine the electrons so as to impinge on gas molecules exiting an orifice. See Plasma source.

Endothermic process   A process that adsorbs energy. Examples: Endothermic chemical reaction; Endothermic phase change.

Endpoint, etching (plasma, semiconductor processing)   The point at which a film has been completely removed as determined by optical emission from the plasma.

Energy   The capacity for doing work.

Energy, kinetic   The energy available due to motion. Example: High-speed ion.

Energy, potential   The energy available due to position or condition. Example: Excited state of an atom.

Engineering notebook   A notebook containing dated entries detailing experiments performed, results obtained, and ideas conceived. For patentable ideas and findings the entries should be read and dated by a non-involved person. Also called a **Laboratory notebook**.

Enthalpy   Heat (energy) content of a system. Example: A high-enthalpy plasma is one that has a high density of energetic particles such as an atmospheric electric arc.

Entropy   A measure of the disorder in a system.

Epitaxial growth (film formation)   Growth of one crystal on another such that the growth of the deposited crystal is determined by the crystalline orientation of the underlying surface.

Epitaxy   Oriented overgrowth of an atomistically deposited film. See Epitaxial growth; Homoepitaxy; Heteroepitaxy.

Epitaxy, heteroepitaxy   Oriented overgrowth on a substrate of a different material or the same material with a different crystalline structure. Example: Silicon on sapphire.

Epitaxy, homoepitaxy   Oriented overgrowth on a substrate of the same material. Example: Silicon on doped silicon.

Equilibrium vapor pressure   The pressure above a surface when there are as many atoms leaving the surface as are returning to the surface (isothermal closed container). See Saturation vapor pressure.

Equivalent weight   The weight of an element or molecule that will combine chemically with 8 grams of oxygen or 1.008 grams of hydrogen. Also called **Combining weight**. Example: Gram equivalent weight.

Ergonomic (furniture)   Designed for comfort and support for a specific type of job to reduce stress and strain on the operator.

Escape depth (characterization)   The depth from which the species to be detected (electron, X-ray, ion) can escape after being created. Example: The low-energy Auger electron created in AES can escape from only a few Ångstrom under the surface of a metal.

Etch-back (pattern)   Generating a thin film pattern by depositing a blanket metallization, then generating a pattern by selective etching, generally using photolithographic processes.

Etch rate (characterization)   The amount of material (mass or thickness) removed per unit time. Often used as a comparative test.

Etch tunnel (barrel etcher)   A tube-shaped grid for shielding the etch region from the rf that sustains the glow discharge in a barrel etcher. The etch tunnel makes the etch region into an afterglow region. See Plasma etcher; Afterglow region.

Etchant   The chemical used for etching.

Etching   The removal of material by chemical reaction to form a soluble or volatile compound.

Etching, cleaning by   Removing surface material (often substrate material) by chemical etching. Removal of the surface material also removes the contamination. See Gross cleaning.

Etching, plasma   Etching in a plasma.

Etching, sputter   Etching a surface by sputtering. Sputter etching is used to clean a surface and also to reveal different crystallographic orientations of the grain structure in the surface.

Etching, vapor   Etching in a chemical vapor.

Etching, wet chemical   Etching in a chemical fluid.

Ethanol (cleaning)   An alcohol that is completely miscible with water and is often used to wipe down vacuum surfaces. See Anhydrous alcohol.

Ethyl alcohol (cleaning)   A non-toxic alcohol derived from grain. Also called **Grain alcohol**. See Denatured alcohol.

Ethylene diamine tetraacetic acid (EDTA) (cleaning)   A chelating agent.

Eutectic composition   A composition that exhibits a local temperature minimum in the solid-liquid boundary in the phase diagram.

Evaporant (PVD technology)   The material to be evaporated.

Evaporation   Vaporization from a liquid surface. See Sublimation.

Evaporation-to-completion (PVD technology) Complete vaporization of the charge of evaporant. A common method of obtaining reproducible film thickness from run-to-run if the geometry of the system and other conditions remain constant.

Evaporation rate, free surface   The amount of material leaving the surface per unit of time when there are no collisions above the surface to cause backscattering of the material to the surface. See Langmuir Equation.

Evaporation source (PVD technology)   The source used to evaporate a material.

Evaporation source, baffle   An evaporation source in which the vapor must collide with several hot surfaces before it can leave the source. Used to evaporate materials such as selenium and silicon monoxide, which vaporize as clusters of atoms or molecules.

Evaporation source, boat   Evaporation from a resistively heated surface in the shape of a boat or canoe.

Evaporation source, coil   A thermal evaporation source in the form of a coil, usually of stranded wire, that is wetted by the molten material and allows deposition in all directions.

Evaporation source, confined vapor   A thermal evaporation source where the vapor is confined in a cavity and the substrate, such as a wire, is passed through the cavity.

Evaporation source, crucible   A container for holding a large amount of molten material. The crucible may be of a number of shapes such as a symmetrical pot or a high-capacity elongated trough (**Hog-trough crucible**).

Evaporation source, e-beam, focused   Evaporation using a focused high-energy low-current electron beam as the means of heating the surface of the material directly.

Evaporation source, e-beam, unfocused   Evaporation using an unfocused low-energy high-current electron beam as the means of heating the material directly or by heating the crucible containing the material.

Evaporation source, feeding   An evaporation source in which the evaporant material is replenished during the deposition process.

Evaporative cooling (vacuum technology)   The cooling of a liquid due to rapid evaporation. In the limit the cooling can actually freeze water in the vacuum system.

Evaporative rate analysis (ERA) (cleaning)   ERA measures the evaporation rate of a radioactive-tagged material that is absorbed by the contaminants on the surface.

Excimer laser   A laser based on a noble gas such as helium or neon where the radiation is from a transition between an excited state and a rapidly dissociating ground state.

Excitation, atomic   The elevation of outer-shell electrons of an atom to a higher energy state. De-excitation gives rise to optical radiation. See De-excitation; Optical radiation; Metastable state.

Exempt solvents (cleaning)   Solvents not subject to pollution regulations. Example: Biodegradable soaps.

Exhaust baffle (vacuum technology)   See Demister.

Exhaust pressure (vacuum technology)   The pressure at the exhaust port of a vacuum pump or in the plumbing from the pumping system to the production environment.

Exhaust system (vacuum technology)   The plumbing system that removes gases and vapors from the work area and is located downstream from the last vacuum pump. This portion of the vacuum system can contain scrubbers to remove undesirable gases and vapors. The exhaust system should not present excessive back-pressure on the vacuum pumping system, particularly during start-up. See Scrubbers; Backpressure.

Exhausted cleaner   A cleaning solution in which the cleaning agents have been depleted to the point that the cleaner is deemed ineffective.

Exoemission (adhesion)   The emission of electrons during fracture. Also called **Fractoemission.**

Exosolution (fluid)   Removal of gases from a fluid generally by reduction of pressure or by heating. Also called **Degassing**.

Exothermic process   A process that releases energy. Examples: Exothermic chemical reaction; Exothermic phase change. See Endothermic process.

Exploding wire, evaporation (film deposition)   The heating and vaporization of a wire by the sudden discharge of an electrical current through the wire and the deposition of the vapor and molten globules thus formed. See Flash evaporation.

External cleaning (cleaning)   Cleaning external to the deposition system.

External processing environment (PVD technology) The processing environment external to the deposition system in which processes such as cleaning, racking, and un-racking take place.

Extinction coefficient (optical)   The optical adsorption per unit path length in a material. Also called **Optical adsorptivity**.

Extra ultra high vacuum (XUHV) (vacuum technology) The pressure range of less than 10^{-9} Torr.

Extractables (cleaning)   Materials that can be extracted from a solid by solvents that it may come into contact with. Example: Extracting phthalates from vinyl gloves by alcohol.

Fab (semiconductor processing)   A production facility, usually for one specific product.

Face mask (contamination control)   Face covering to prevent contamination from fluids from the mouth or nose, or particulates from the face or facial hair.

Face mask (safety)   Face covering to prevent chemicals from coming into contact with the face.

Fahrenheit temperature scale   A temperature scale based on the freezing point of water being 32°F and the boiling point of water being 212°F under standard pressure conditions. See Temperature scale.

Fail-safe design (vacuum technology)   A design such that the system will assume a safe and non-contaminating configuration if there is a mechanical, electrical, or coolant failure. See What-if game.

Failure analysis (adhesion)   The analysis of the failed interface and other contributing factors to try to determine the cause of the failure.

Faraday's Law of Electrolysis   Faraday's Law of Electrolysis states that the amount of material dissolved or deposited in an electrolysis cell is proportional to the total charge passed through the cell.

Fatigue   Reduction of some property of a material after some period of stress.

Fatigue, chemical   Fatigue after exposure to a chemical environment. Example: Reduction in strength due to stress corrosion.

Fatigue, mechanical   Fatigue under mechanical motion, deformation, etc. Example: Workhardening (reducing the ductility of a metal).

Fatigue, static   Fatigue due a continuously applied stress with no motion. Example: Static fatigue failure in glass.

Feedback (process)   The control of the output of a process by the return of information about the output to the input.

Feeding source (evaporation)   An evaporation source in which the evaporant material is replenished during the deposition process. See Evaporation source; Flash evaporation.

Feeding source, pellet   A mechanism to feed individual pellets into a molten pool to replenish the charge or onto a hot surface for flash evaporation.

Feeding source, powder   A mechanism to feed powder into a molten pool to replenish the charge or onto a hot surface for flash evaporation.

Feeding source, rod-feed   A focused e-beam source where the surface of the end of a rod is heated and the molten material is contained in a cavity of the rod material. As the material is vaporized, the rod is moved so as to keep the molten material in the same position with respect to the e-beam.

Feeding source, tape feed   An evaporation source where the melt material is continually or periodically renewed by a tape being fed into the molten material. Generally a tape is easier to feed than a wire.

Feeding source, wire feed   An evaporation source where the melt material is continually or periodically renewed by a wire being fed into the molten material.

Feedthrough (vacuum technology)   A device for transmitting electrical, optical, or mechanical signals or fluids through the wall of a vacuum chamber. The feedthrough is generally mounted on a flange. See Flange.

Feedthrough, electrical   A feedthrough that allows passage of electrical signals into the deposition chamber.

Feedthrough, fluid   A feedthrough that allows passage of fluids into the deposition chamber. The fluid may be hot or cold, even to cryogenic temperatures.

Feedthrough, magnetic   A feedthrough that allows passage of magnetic flux into the deposition chamber. Also called a **Magnetic window**.

Feedthrough, mechanical   A feedthrough that allows passage of mechanical motion into the deposition chamber. The vacuum sealing may be by: a single "O" ring, differentially pumped "O" rings, Ferrofluidic seals, a rotary magnetic drive through a solid metal wall, or a wobble motion using a bellows to give a rotary motion in the chamber.

Feedthrough, optical   A feedthrough that allows passage of optical signals into or out of the deposition chamber. Also called a **Window**.

Ferric oxide (Fe_2O_3)   A polishing compound. Also called **Jeweler's rouge**; **Red ochre**. See Cerium oxide (CeO_2).

Ferrofluid (sealing)   A colloidal suspension of colloidal (approx. 10 nm) magnetic particles (Fe_3O_4) in a fluid. See Ferrofluidic seal; Liquid "O" ring.

Ferrofluidic seal (vacuum technology)   A rotary motion feedthrough that is sealed by an oil-based ferrofluid held in place by a magnetic field.

Ferromagnetic material   Material in which the electron spins can be preferentially oriented to produce a permanent magnetic moment even when there is no externally applied magnetic field.

Field emission, electron   Emission of electrons under a high-electric field, usually from a point.

Field emission, ion   Creation of gaseous ions in a high-electric field by the tunneling of electrons from the gaseous atoms to a surface.

Field emission, ion, liquid metal   Creation of metal ions by evaporation from a liquid metal wetted point in a high-electric field.

Field emitter tip   Sharp point used to generate electrons or metal ions by high-electric-field effects.

Field evaporation   Vaporization from a sharp tip due to a high electric field.

Field-free region (plasma)   A region in which there is no electric field. Usually generated by having the region surrounded by an electrical conductor (solid or as a grid).

Film (substrate)   A free-standing flexible structure of limited thickness. Also called a **Web**.

Film ions (PVD technology)   Ions of the condensable film material being deposited. Often accelerated to a high kinetic energy in an electric field.

Filtered arc source   An arc vaporization source designed to filter out the macros, generally by deflecting the plasma beam. See Arc source; Plasma duct.

Filtration (cleaning)   Removal of a species from a fluid.

Filtration, microfiltration   Removal of particles of 0.1 to 10 microns.

Filtration, particle   Removal of particles having a size of 1 to 100 microns.

Filtration, reverse osmosis (RO)   A method of removal of ionic-sized particles by the use of a membrane filter.

Filtration, ultrafiltration   Removal of particles of 0.001 to 0.1 micron.

Fin (ceramic)   A thin edge formed on a ceramic during the fabrication process. Much the same as a burr except not due to deformation. See Burr.

Final rinse (cleaning)   In wet cleaning the surface being cleaned should be kept wet until the final rinse—that is, the last rinse before drying. This rinse should be done with ultrapure water to a specified resistivity to minimize residues. See Rinse-to-resistivity.

Fines   Particles smaller than the average or specified particle size. See Mesh sizing.

Finger cots (cleaning)   Coverings, usually of rubber, that only cover the tips of the fingers and can be used instead of gloves when handling material in some cases. Can be used inside cloth gloves.

Fire side (glass)   The side of the glass from a float glass plant that has not been in contact with the molten tin. See Tin side.

First surface (optical)   The surface of the optical substrate facing the incident radiation. Example: First surface mirror that is metallized on the "frontside" of the glass. See Second surface.

Fisheye (defect)   A flow defect in a flow-coated surface resulting from a particulate or an inclusion on the surface.

Fixture (film deposition)   The removable and generally reusable structure that holds the substrates during the deposition process. The fixture is generally moved, often on several axes, by tooling during the deposition process. In some cases the same fixture is used to hold the substrates during the cleaning process. See Rack; Tooling.

Fixture, cage (film deposition, electroplating)   A container with wire mesh sides that contains loose parts and is rotated during the deposition process to allow complete coverage of the parts. Also called a **Barrel fixture**.

Fixture, callote   A hemispherical cap-shaped fixture on which the substrates are mounted. Often used in thermal evaporation to keep the substrate surfaces an equal distance from the point-evaporation source and to keep the angle-of-incidence of the deposition normal to the substrate surfaces.

Fixture, carousel   A fixture on which parts are mounted, then moved in a circular motion (like a merry-go-round). Example: In front of a sputtering target or between two sputtering targets.

Fixture, cassette (semiconductor processing)   A storage fixture that holds wafers so that the paddle can perform a **Pick-n-place** motion. See Paddle.

Fixture, Christmas tree   A fixture that has a number of branches on which parts are hung. Also called a **Tree fixture**.

Fixture, drum   A cylindrical fixture where the substrates are mounted on the walls of a cylinder or mounted on structural members positioned in a cylindrical arrangement.

Fixture, drum, rotisserie   A planetary arrangement using a cylindrical drum fixture where the parts are mounted in a cylindrical arrangement around the axis of rotation of the drum and rotate about a second axis.

Fixture, ladder (thermal evaporation)   A fixture for holding a number of evaporator filaments in a vertical array so as to approximate a line source.

Fixture, pallet   A planar surface on which the substrates lie or are mounted. The pallet may be held horizontally or vertically. Often the initial angle-of-incidence of the depositing material is high, which can lead to film-density problems.

Fixture, planetary A fixture that has a motion around one fixed axis and several moving axes in a plane.

Fixture, vibratory pan A fixture for coating small parts by placing them in a pan that is vibrated, causing the parts to move about and allowing 100% coverage of the part. Also called a **Shaker table**.

Flakes (contamination control) Particles of film material that become dislodged in the vacuum system and generate particulate contamination in the system.

Flame spray (thermal spray) Melting small particles in a flame, such as an oxygen-acetylene torch, accelerating the molten particles in a high-velocity gas stream (1200 ft/sec), and "splat cooling" them onto a surface.

Flame treatment (polymer) A method of oxidizing the surface of a polymer web to increase its surface energy by subjecting it to a flame in air. See Corona treatment; Plasma treatment.

Flammable gas A gas that is flammable in a mixture of 13% or less (by volume) with air. See Flash point.

Flange (vacuum technology) A mechanical structure designed to allow sealing of one structure to another, usually to isolate vacuum from the ambient pressure. The flange may provide sealing by use of an **Elastomer seal**, a **Deformation seal**, or a **Shear seal**. Often feedthroughs are mounted on the flange. See Feedthrough.

Flange, blank-off (vacuum technology) A flange that does not contain a feedthrough or other component that is used to seal a port.

Flange, female (vacuum technology) A flange with a recessed sealing feature designed to seal to a male flange.

Flange, male (vacuum technology) A flange with a protruding sealing feature designed to seal to a female flange.

Flange, rotatable (vacuum technology) A flange that can be rotated to align the bolt-holes in any position.

Flange, sexless (vacuum technology) A flange whose mate has an identical sealing structure.

Flash (electroplating) A very thin coating (40 millionths of an inch [1 micron] or less) deposited by electroplating. Often used to prevent corrosion of a surface. Example: Flash of gold. See Strike.

Flash deburring The burning off of a burr in a flame front produced by an explosion.

Flash evaporation (film deposition) The deposition of a material by rapid heating so that there is no time for diffusion or selective evaporation. Flash evaporation is used to deposit alloy materials where widely different vapor pressures prevent uniform thermal vaporization of the elemental components of the alloy.

Flash evaporation, exploding wire The heating and vaporization of a wire by the sudden discharge of an electrical current through the wire.

Flash evaporation, laser ablation Vaporization of a surface by the adsorption of energy from a laser pulse.

Flash evaporation, pellet feed Where individual pellets are fed onto a hot surface and are completely vaporized before the next pellet is dropped.

Flash evaporation, wire tapping Where the tip of a wire is periodically tapped against a hot surface so the tip of the wire is periodically vaporized.

Flash point (safety) The lowest temperature at which vapors will ignite and burn when exposed to an ignition source. Important consideration when using flammable materials.

Flash rust (cleaning) The oxide (rust) layer that rapidly forms on the dry, oxide-free surface of steel.

Flashover, surface (electrical) Electrical discharge across the surface of an insulator. See Vacuum breakdown (arc).

Flaws, interfacial (adhesion) Flaws in the interfacial material, such as cracks and voids, that concentrate stress and provide initiation points for fracture. Their presence lowers the fracture toughness of the interfacial material. See Flaws, surface.

Flaws, surface (substrate, adhesion) Flaws in the substrate surface such as cracks or voids that become incorporated into the interfacial region. Their presence lowers the fracture toughness of the interfacial material.

Flint glass (optical) A high-dispersion, relatively high-index glass used in diverging lens elements.

Flip-chip bonding (semiconductor processing) When the circuit die is connected directly to the printed circuit board or chip carrier by means of solder bumps. See Die.

Float glass (substrate) Glass sheet formed by continuously pouring molten glass onto a bed of molten tin. Most window glass is made by this technique, which leaves a layer of tin oxide on one surface. Typical composition of float glass is SiO_2 = 72-74 wt%, Na_2O = 12-15, CaO = 6-10, MgO = 3-5, and Al_2O_3 = 0.2-1.5.

Floating potential The electrical potential assumed by a material that is electrically isolated from ground.

Flocculate (cleaning) To cause to come together into a mass. Flocculation is performed on turbid water before the purification operation. See Flocculating agent.

Flocculating agent (cleaning) Agent used to cause small particles to coalesce into a large mass. Also called a **Flocculant**. Example: Used in water treatment prior to filtration.

Flood panel (vacuum technology) A water-cooled double-walled panel, such as the wall of a vacuum chamber, that is used to remove process heat from the surface.

Flow, laminar (cleaning) A streamline gas or fluid flow without turbulence.

Flow, mass (vacuum technology) Particles per second

passing by a position. Also called **Mass throughput**.

Flow, molecular (vacuum technology)   Gas flow conditions where there are few collisions between molecules because of the long mean free path for collision (low pressure).

Flow, transition (vacuum technology)   Gas flow conditions intermediate between viscous flow and molecular flow where the flow characteristics are determined by molecular collisions and collisions with the walls of the duct.

Flow, turbulent   A gas or fluid flow where local velocities fluctuate in an irregular and random manner. See Velocity.

Flow, viscous (vacuum technology)   Gas flow conditions where the mean free path for collision is very small compared to the dimensions of the system.

Flow chart, process (manufacturing)   A schematic diagram of the processing, including inspection, characterization, handling, and storage, that a substrate encounters in going from the as-received material to the final product. The flow chart is useful in determining that complete documentation has been developed for all phases of the processing.

Flow coating (PVD technology)   Coating by flowing a fluid (lacquer) over a surface, then letting the fluid harden by evaporating a solvent or by heating. Used to apply basecoat material, particularly for producing a smooth surface. Also used to apply topcoat films. See Dip coating.

Fluid application (cleaning)   The various means of applying a cleaning or rinsing solution to a surface in order to clean or rinse it.

Fluid application, immersion   To leave in a cleaning solution for a period of time, often with mechanical movement of the part and agitation of the solution. Also called **Soaking**.

Fluid application, spray   Spraying with a cleaning or rinsing agent with a **Low-pressure spray** (< 100 psi) or a **High-pressure spray** (> 1000 psi).

Fluid application, ultrasonic   Cleaning or rinsing using the jetting action of the collapse of cavitation bubbles in contact with a surface to provide agitation. Frequencies in the range of 20 kHz to 100 kHz.

Fluid ounce   A measure of fluid volume equal to $1/32$ of a quart. Often just called an ounce.

Fluidized bed   A body of powder that is kept in motion by a flow of gas and/or vibration. Particles immersed in the bed can be coated or polished. See Pack cementation; Vibratory polishing.

Fluorophores   Fluorescent materials.

Flux (particle bombardment)   The number of particles per unit area per unit time. Example: Ions per cm^2 per second. Also called the **Dose rate**.

Flux distribution (film deposition)   The angular distribution of the particles incident on the substrate surface.

Flux distribution (vaporization)   The angular distribution of the particles leaving a vaporization source. See Cosine distribution.

Flux ratio (ion plating)   The ratio of the number of energetic bombarding particles to the deposition rate of the depositing condensable film atoms.

Fluxing (cleaning)   A metal-cleaning technique that operates by dissolving or floating off the oxides on a surface using a hot molten fluid solvent, which is often a borate.

Fogger (cleaning)   Machine for generating fine particles for checking mechanical filters in an air circulation system. See Dioctyl phthalate (DOP).

Footprint (equipment)   The amount of floor space that a piece of equipment occupies.

Forcefill (metallization, semiconductor processing)   The use of a high isostatic pressure (~60 Mpa) and temperature (~400°C) to close voids in thin film aluminum metallization.

Foreline (vacuum technology)   The plumbing between a high-vacuum pump and its backing pump.

Foreline pressure (vacuum technology)   The pressure in the foreline at the outlet of the high vacuum pump.

Forepump (vacuum technology)   A vacuum pump used to keep the discharge pressure of a high vacuum pump below some critical value. The forepump may also be used as a roughing pump by proper valve sequencing. Also called a **Backing pump**. See Roughing pump.

Forming gas   A gas mixture of nitrogen and hydrogen (usually 90 : 10) that has a low flammability.

Fourier transform infrared (FT-IR) analysis (characterization)   Infrared spectroscopy using the adsorption of infrared radiation by the molecular bonds to identify the bond types that can absorb energy by vibrating and rotating. In Fourier transform infrared spectrometry (FT-IR) the need for a mechanical slit is eliminated by frequency modulating one beam and using interferometry to choose the infrared band.

Fractional distillation   A means of purifying a material by selective vaporization of the more volatile material(s). Purification may be of the material remaining or of the material volatilized. Used to purify evaporant materials (vacuum evaporation), solvent cleaners, and pump oils.

Fractionation, by evaporation (PVD technology)   When preferential vaporization of one constituent of a vaporizing melt occurs due to its higher vapor pressure, leaving the melt with an increasingly higher proportion of the less-volatile material. See Fractional distillation;

Raoult's Law.

Fractionation, gas, by pumping (vacuum technology) Changes in the composition of gas in a vacuum chamber due to preferential pumping of one gas species over another. Example: Cryopumping increases the relative helium content in the chamber since it pumps helium poorly.

Fractoemission (adhesion) The emission of electrons during the fracture of a dielectric, brittle solid due to charge separation and arcing. Also called **Exoemission**.

Fractograph The picture of a fractured surface.

Fracture (adhesion) The generation of two free surfaces through the bulk of a material or at an interface between materials.

Fracture initiation (adhesion) The starting point of a fracture. Often fracture, particularly in a brittle material, starts at a flaw or point of stress concentration. The amount of stress that must be imposed to initiate a fracture when there is no flaw present. The stress needed to initiate a fracture is usually much higher than that needed to propagate the fracture.

Fracture propagation (adhesion) The extension of a fracture through the material.

Fracture toughness (adhesion) A measure of the amount of energy needed to cause fracture propagation.

Fragment pattern (mass spectrometry) The portion of the spectra from a mass spectrometer due to the breaking up of complex molecules by electron bombardment. Also called a **Cracking pattern**. See Fragmentation.

Fragmentation (plasma technology) Breaking up a molecular species into less complex species.

Frank-van der Merwe growth mode (film formation) Layer-by-layer growth where there is strong interaction between the depositing atoms and the substrate. Complete coverage of the substrate is attained in a few monolayer film thickness. See Volmer-Weber (island) growth; Stranski-Krastanov (pseudomorphic) growth.

Free energy, surface The energy per unit surface area that results from the asymmetrical bonding of the surface atoms. See Surface tension.

Free-machining alloys (metallurgy) Alloys that have additions of sulfur or selenium to make them more easily machinable. These additions can cause trouble later in passivation and corrosion protection.

Free-span (web coating) A web coating machine in which the web is not in contact with a surface during the actual film deposition part of the web coating process. See Chill drum.

Freeboard ratio The ratio of the height of the freeboard above the vapor level, to the closest horizontal liquid dimension, in an old-style vapor degreaser.

Fretting wear (contamination control) A type of wear where adhesion between two contacting surfaces in relative motion causes the wear.

Friction (vacuum technology) The resistance of surfaces in contact to move relative to each other. The higher the friction, the more likely the galling and the generation of particulate contamination. See Coefficient of friction.

Frictional drag (vacuum technology) The deceleration force applied to a moving surface by a gaseous environment in contact with the surface. See Molecular drag pump.

Frontend (semiconductor processing) Equipment and processes that are used to fabricate a wafer. Examples: Ion implantation machine; PECVD equipment; Chemical-mechanical polishing (CMP) equipment. See Back-end.

Frost (vacuum technology) The solid condensed material that forms on cold surfaces and reduces the thermal conduction from the cold surface to the surface of the frost. The frost is removed by **Regeneration.**

Full flow (leak detection) When all of the helium passes through the leak detector whose pump has replaced the backing pump of the vacuum system.

Full flow (pumping) When there is no conductance-reducing component deliberately placed in the path of the gas. See Throttled flow.

Functionalization (surface preparation) Generation of radicals or dangling bonds on the surface of a polymer to increase the surface reactivity.

Functional coating A coating that improves the functional properties of a surface such as wear-resistance, corrosion-resistance, abrasion-resistance, bondability, etc.

Fused salt electrodeposition (electrodeposition) Electrodeposition using a fused salt, such as a chloride or a fluoride, as the electrolyte instead of an aqueous electrolyte. See Metalliding.

Fused salt metalliding (electrodeposition) Deposition of a film or coating using fused salt electrodeposition. Often the deposited material reacts extensively with the substrate surface, forming an alloy or a compound. See Fused salt electrodeposition; Metalliding.

Fusion coatings Coating a surface by fusion of the additive material to the surface. Examples: Enameling; Thick film metallizing. See Thick film metallization.

G

Galling (contamination control) Surface damage due to adhesion and fracturing of surfaces in contact. Galling is a source of particulate contamination in vacuum systems containing moving parts.

Galvanize The process of depositing zinc on a surface, usually by hot dipping or electroplating.

Galvanic corrosion Electrochemical corrosion due to the voltage generated by dissimilar metals in contact with an electrolyte present. Examples: Galvanic corrosion between a film matrix and a precipitated phase (Al_2Cu in Al metallization), Chromium carbide in an alloy matrix in stainless steel weldments.

Garnet   A naturally occurring abrasive material that is composed of metal silicates.

Gas   A state of matter in which the molecular constituents move freely and expand to fill the container that holds it. Generally the term includes vapors. See Vapor.

Gas, ideal   A gas that is composed of atoms and molecules that physically collide but otherwise do not interact. Low-pressure gases are generally treated as **Ideal gases**.

Gas, non-ideal   A gas that does not obey the Ideal Gas Law because of atomic and molecular interactions other than physical collision. Example: Water vapor at room temperature. Also called a **Real gas**.

Gas ballasting (vacuum technology)   The introduction of a non-condensing gas into the compression stage of a vacuum pump to dilute the vapors in the pump so that they will not be condensed by compression above their saturation vapor pressures.

Gas blanket   A protective environment formed by an inert gas surrounding the surface.

Gas cabinet (gas distribution)   A storage enclosure designed to provide a controlled local environment to a gas cylinder and to provide safety conditions where needed.

Gas conversion   Forming a hard diffusion layer by heating a surface in contact with a reactive gas that can react with a constituent of the alloy to form a dispersion strengthened layer (**Case**). Example: Gas nitridation.

Gas discharge   See Glow discharge.

Gas evaporation   Vaporization into a gaseous environment that has a gas density sufficient to allow collisions that lead to gas phase nucleation and the generation of ultrafine particles in the gas. See Ultrafine particles.

Gas incorporation (film formation)   Incorporation of soluble or insoluble gases during film growth, either by physical trapping or by low-energy implantation by bombarding species. Example: Incorporation of helium in gold films. See Charging, hydrogen.

Gas scatter plating (film deposition)   Increasing the throwing power of the depositing atoms by scattering the atoms in a gaseous atmosphere. Does not work very well without a plasma due to gas phase nucleation and the deposition of ultrafine particles. When a plasma is present the ultrafine particles become negatively charged and do not deposit on the substrate, particularly if the substrate is at a negative potential (as in ion plating).

Gas scattering   Scattering of a high-velocity atom by collision with gas molecules. See Mean free path; Ther-

malization; Gas scatter plating.

Gas-phase nucleation (particle formation)   The nucleation of atoms in a gaseous environment where multibody collisions allow the removal of the energy released on condensation. See Gas evaporation.

Gaseous arc   An arc formed in a chamber containing enough gaseous species to aid in establishing and maintaining the arc. See Vacuum arc.

Gasket (vacuum technology)   The object between sealing flanges that deforms or shears, thus creating the vacuum-tight seal. See Flange.

Gate valve (vacuum technology)   A mechanical sealing valve where the motion of the sealing plate is mostly parallel to the plane of the seal. Generally the valve opening is round so that the maximum opening is achieved with the use of the least sealing area. See Vacuum valves.

Gauge   A measuring device. Example: Vacuum gauge. See Sensors.

Gauge   A thickness unit. Example: 18 gauge steel sheet.

Gauge   A diameter unit. Example: 12 gauge electrical wire.

Gauge band (web coating)   A continuous lane of film in the machine direction of the roll that is abnormally thick (**Hard band**) or thin (**Soft band**).

Gauss   Unit of magnetic field intensity equal to one Maxwell/cm^2 or 10^{-4} Weber/m^2. See Oersted (cgs system); Tesla (SI system).

Getter (vacuum technology)   A material that will react with or adsorb reactive gases in the vacuum environment.

Getter (vacuum technology)   To remove gases either by a chemical reaction so as to form non-volatile solid species containing the gas, or by absorption of the gases in the getter material.

Getter pump (vacuum technology)   A vacuum pump that operates by reaction of a surface with the gaseous species to form a non-volatile reaction product or by absorption of the gases into the bulk of a getter material. In reaction-type getter pumps the getter materials are often deposited by evaporation or sublimation. Adsorption-type getter pumps are sometimes called **Non-evaporative getter pumps**. See Vacuum pump.

Getter pumping, during deposition (PVD technology)   The gettering action (pumping) of reactive gases that accompanies the deposition of a reactive film material such as titanium in an oxygen environment.

Gilding   Overlaying a surface with a very thin free-standing film (e.g., gold or silver) that is adhesively bonded to the surface or held to the surface by electrostatic forces.

Gilding, depletion   The leaching of base metals from a gold alloy to form a gold-rich surface that can be burnished to a high density and luster. See Leaching.

Glass (substrate)   A non-crystalline (amorphous) material. Common inorganic glasses are composed of a mixture of oxides and additives (**Glass formers**) that inhibit crystallization. There also are **Metallic glasses** and **Organic glasses**. See Insulated glass; Laminated glass; Tempered glass; Hardened glass.

Glass, aluminosilicate   A high-melting-point glass composed of a mixture of aluminum oxide and silicon oxide.

Glass, float   Glass sheet formed by continuously pouring molten glass on a bed of molten tin. Most window glass is made by this technique, which leaves a layer of tin oxide on one surface.

Glass, hardened   Glass that has been strengthened by heat treatment by a factor of two or so but not to the level of tempered (fully tempered) glass.

Glass, high lead   A low-melting-point, high-index of refraction glass that contains a high percentage of lead oxide.

Glass, machine drawn   Flat glass sheet formed by drawing. See Float glass.

Glass, soda-lime   A common glass made by fusion of sand with sodium carbonate or sodium sulfate and lime or limestone.

Glass, stressed (substrate)   Glass in which the surface has been put into compressive stress to strengthen the glass by making the generation and propagation of a surface flaw more difficult. The compressive surface region can be generated by **Thermal quenching** or by **Ion substitution**.

Glass, tempered   Glass that has a high compressive stress on the surfaces and a high tensile stress at the midplane. When fractured the tempered glass breaks up into small shards. Also called **Toughened glass**.

Glass bead blasting (cleaning)   Grit blasting using glass beads. See Shard.

Glass formers   Materials that are added to glass formulation to help keep the composition from crystallizing. Examples: PbO; CdO; Bi_2O_3.

Glass transition temperature   The temperature above which a brittle glassy material (polymer, oxide glass, etc.) becomes ductile. Also called the **Strain point**.

Glaze (coating)   A smooth, glassy coating formed by firing a glass frit on a surface. See Thick film.

Glaze (wear)   A smooth surface formed by sliding. See Burnishing.

Glazier   A person who installs glass, usually in windows.

Glazing   A transparent or translucent material (glass or plastic) used to admit light and/or to reduce heat loss; used for building windows, skylights, or greenhouses, or for covering the aperture of a solar collector.

Global warming potential (GWP) (cleaning)   A rating for the potential of a vapor to contribute to global warming. See Ozone depletion potential (ODP).

Glove box   A controlled-atmosphere box where handling is done with gloves that extend through hermetic seals into the box. The atmosphere in the glove box may be made inert using nitrogen or argon instead of air or may have a very low moisture content (**Dry box**). Also called **Isolators** (England).

Gloves (cleaning)   Hand covering that comes into contact with substrates or fixtures and solvents. The gloves should have low-extractables as far as the solvents are concerned. See Finger cots.

Glow (plasma)   The visual emission from a glow discharge, particularly the plasma region.

Glow bar (PVD technology)   A high-voltage electrode that allows a glow discharge to be established in a vacuum chamber for cleaning and surface-treatment purposes. The glow bar should be as large as possible in order to generate as uniform a plasma as possible throughout the chamber.

Glow discharge (plasma)   The plasma generation region and other contiguous plasma-containing regions such as the plasma region, the afterglow region, and the wall sheath. Also called a Gas Discharge.

Glow discharge cleaning   Subjecting a surface to a plasma of an inert or reactive gas to enhance desorption of gases and, in the case of reactive gas plasma, by forming volatile species that leave the surface. Cleaning occurs by the action of ions accelerated across the wall sheath, radiation from the plasma, and energy released on the surface by the recombination of ions and electrons. In the cases of reactive gas plasmas, chemical reactions occurs on the surface. See Ion scrubbing; Reactive plasma cleaning.

Glow discharge mass spectrometry (GDMS)   An analytical technique where atoms are sputtered from a surface, ionized in the plasma, then mass analyzed in a mass spectrometer.

Glue-layer (adhesion)   An intermediate layer between the film and the substrate used to increase adhesion. Also called a **Bond coat**. Example: The titanium layer in a titanium-gold metallization on an oxide. The titanium chemically reacts with the oxide and alloys with the gold.

Gold black   Ultrafine particles of gold, often made by gas evaporation. Used as an infrared radiation adsorber in bolometers.

Gold-filled   Gold layer is mechanically bonded (cladded) to the surface by rolling, soldering, or drawing. Gold-electroplated or gold-PVD-coated items cannot legally be called gold-filled.

Goniometer, contact angle (cleaning, surface treatment)   An instrument for measuring the angle-of-contact of a fluid with a surface using direct observation or projection techniques. See Contact angle.

Gowning protocol (contamination control)   The carefully choreographed moves for putting on (donning) cleanroom clothing (head covering, face covering, bunny suits, booties, and gloves) to minimize contamination of the outer surface of the clothing.

Graded interface (film formation)   When the interfacial region between a film and a substrate has composition or properties that vary throughout the thickness. See Interphase material.

Grading   The gradual changing of a property or composition from one value to another. Examples: Graded density coating; Grading composition from Ti to TiN by controlling nitrogen availability during reactive deposition; Grading TCE by grading glass composition in a glass-to-metal seal.

Grain (gr) (weight)   The smallest unit of weight in the avoirdupois weight system. 1 grain = 0.0648 grams.

Grain (crystallography)   A volume of material having a specific crystalline composition or a different orientation with respect to its neighboring grains.

Grain boundary   The boundary between two crystalline regions that have different grain orientations.

Gram (g) (weight)   A unit of weight.

Gram equivalent weight   The gram molecular weight divided by the valence of the ion of interest. Example: The gram equivalent weight of carbon in the +4 valence state is 3 grams (i.e., 12 divided by 4). See Normal solution.

Gram molecular (or atomic) weight   The weight of a compound (or element) in grams. Example: 12 grams of CO_2. See Mole; Molar solution.

Green cleaning (cleaning)   Cleaning using environmentally benign chemicals and processes.

Grit (cleaning)   A particulate material used in abrasive cleaning and surface roughening. Examples: Steel shot; Fractured cast iron shot; Silica sand (sandblasting); Alumina; Magnesium carbonate. See Grit size.

Grit blasting (cleaning)   Removal of surface material (gross cleaning) or roughening a surface by entraining grit in a high-velocity gas stream directed onto the surface. See Glass bead blasting.

Grit size (cleaning)   A measure of the particle size and size distribution used in abrasive cleaning or grit blasting. Example: 120 grit cast iron grit. See Mesh sizing.

Gross cleaning (cleaning)   Cleaning by removal of surface material as well as contaminant material. See Specific cleaning.

Ground (electrical)   The electrical plane, usually earth, that has a common zero potential and to which most electrical circuits are referenced by being attached (i.e., grounded).

Ground loop (electrical)   The condition by which an electrical circuit is not attached directly to ground but rather goes through another piece of equipment that prevents the electrical circuit from being referenced to the ground (zero) potential.

Ground shield (plasma technology)   A grounded surface placed at less than a dark-space distance from a DC cathode surface in order to prevent a glow discharge from forming on the surface. See Paschen curve.

Gusset (vacuum technology)   A rib used to strengthen a plate to prevent it from bending under pressure.

Gyro radius (plasma)   The radius of the path that an electron takes in a magnetic field. See Larmour radius.

H

Hall effect   The development of a transverse electric field in a current-carrying conductor placed in a magnetic field. Hall-effect probes are used to measure magnetic field intensities. See EXB drift.

Halogenated solvents (cleaning)   Solvents containing the halogens (Cl, Fl, Br). See Chlorofluorocarbon.

Hard arc   High-current, sustained arc usually due to shorting (by a flake) or thermoelectron emission from a heated oxide particle on a high-voltage electrode.

Hard coating   A coating that extends the life of a tool that is subject to wear such as a drill bit, an extrusion die, an injection mold, etc. The mechanism may not be entirely related to hardness of the coating. For example, the coating can reduce the friction and thus prolong tool life or it may provide a diffusion barrier that prevents adhesion and galling.

Hard vacuum (vacuum technology)   See High vacuum.

Hard water (cleaning)   Water containing dissolved ions (e.g., Ca, Fe, Mn) that can leave a residue if evaporated or if they react with other chemicals such as phosphates to form water-insoluble compounds. See Soft water; Deionized water.

Hardness   The resistance of a surface to deformation. Generally measured by the resistance to indentation.

Haze (cleaning)   Surface morphology that gives diffuse reflection from an otherwise smooth (specular) surface. Example: Haze on a float glass surface from a residue of tin oxide.

Hearth (e-beam evaporation)   The water-cooled structure that has a depression called a **pocket** in which the material to be evaporated is contained. See Pocket; Skull; Liner, pocket.

Heat affected zone (HAZ)   The region near a weld joint that is affected by heating during the joining process. Example: The HAZ in high-carbon stainless steel that has been welded contains precipitated chromium carbide that can cause problems with galvanic corrosion.

Heat exchanger A high-surface-area device to maximize the heat exchange between two physically separate gas or liquid materials.

Heat mirror A thin film structure that transmits the visible spectrum while reflecting the near-infrared.

Heat of condensation Heat released by the physisorption or chemisorption of species on a surface. See Heat of vaporization.

Heat of reaction Heat taken up (endothermic) or released (exothermic) during a chemical reaction.

Heat of solution (safety) Heat released or taken up during solution. Example: Add acid slowly to water to prevent local heating and splattering.

Heat of vaporization Heat taken up during the vaporization of a molecule from a surface and released on condensation. Example: The heat of vaporization of gold from a tungsten surface equals about 3 eV per atom. See Heat of condensation.

Heat-strengthened glass Glass that has been strengthened by creating a stress profile that is compressive at the surface and tensile in the center. Strength is more than (2X) annealed glass but much less than fully tempered glass.

Heating mantle (vacuum technology) A heating device that conforms to the shape of the vacuum chamber (system) and that is used for baking out the system. See Bake-out.

Helicon plasma source (plasma) A plasma source in which microwave power is used to accelerate electrons in a gas in the presence of a constant magnetic field. See Plasma source.

Helium (leak detection) Gas (amu = 4) used for "helium leak detection."

Helium leak detector (vacuum technology) A mass spectrometer tuned to the helium peak that is attached to a vacuum chamber and monitors any change in helium concentration in the chamber as helium gas is directed toward the exterior of the chamber.

Hermetic seal (vacuum technology) An air-tight seal.

Heteroepitaxy Oriented overgrowth on a substrate of a different material or the same material with a different crystalline structure. Example: Silicon on sapphire. See Homoepitaxy.

Heterogeneous nucleation (film formation) Nucleation of one material on a different material. Example: Silicon on sapphire. See Homogeneous nucleation.

Hideouts (cleaning) Areas on a surface that are difficult to clean such as cavities, pores, or surfaces in close contact.

High-efficiency particle air (HEPA) filter (contamination control) See Mechanical filter.

High-energy neutrals (sputtering) High-energy neutral species formed by neutralization and reflection of the high-energy bombarding ions during sputtering.

High-energy neutrals (plasma chemistry) High-energy neutral species formed by a charge exchange process between a high-energy ion and a slow atom.

High solids content (polymer coating) Having a low content of volatile components such a volatile organic compounds (VOCs) in the coating material. See Volatile organic compound (VOC).

High vacuum (vacuum technology) A gas pressure where there is molecular flow, a low particle density, and a long mean free path for gas phase collisions. Generally taken as a pressure below about 10^{-5} Torr.

High vacuum (PVD technology) A gas pressure in which there is no significant amount of gaseous contamination that will affect the deposition process or the properties of the deposited film.

High-vacuum pump (vacuum technology) A device for producing a high vacuum, either by capturing and holding the gases or by compressing and expelling the gases. See Vacuum pump.

High-velocity-oxygen-fuel (HVOF) spray A thermal spray process where oxygen and fuel gases at high pressures and flow rates are burned in a combustion chamber and particles of the coating material are injected into the expanding gases as they flow through a nozzle where they are melted and achieve supersonic speed.

Hillock (metallization) A raised mound (bump) on a metallization film, often formed on ductile metals during electromigration or when there is a high compressive film stress. See Bleb.

History, of materials (substrates, cleaning) The history of a material includes specification of raw materials, fabrication techniques, storage times and environments, etc. In many cases the history of the material to be coated determines what must be done to clean or prepare the surface. In addition, changes in the history from lot-to-lot can be an unacceptable process variable. See Outgassing; Outdiffusion.

Holding pump (vacuum technology) A small-capacity pump used to maintain the foreline pressure of certain types of high-vacuum pumps when the use of the main backing pump is not justified. See Backing pumps.

Holidays (electroplating) Voids in the interface between two materials.

Hollow cathode (plasma) A cathode with a deep cylindrical cavity or tube such that the electrons are trapped in the cavity and are effective in ionizing gases in the cavity. The cathode can be heated to the point that there is thermoelectron emission (**Hot hollow cathode**). The hollow cathode can be used as an electron source.

Hollow cathode discharge (HCD) lamp A light source

using a hollow cathode discharge whose emission spectrum is characteristic of the material of which the cathode is comprised.

Homoepitaxy   Oriented overgrowth of a film on a substrate of the same material. Example: Silicon on doped silicon. Also called **Isoepitaxy**. See Heteroepitaxy.

Homogeneous nucleation   Nucleation of atoms on a surface of the same material. Example: Silicon-on-silicon. See Heterogeneous nucleation.

Hood test (vacuum technology)   Placing a vacuum chamber in a bag filled with helium to measure the total real leak rate into the chamber. Also called a **Bag check**.

Hot cathode ionization gauge (vacuum technology)   An ionization vacuum gauge in which the electrons for ionization are obtained from a thermoelectron emitting filament. See Vacuum gauge.

Hot dip galvanizing   Coating of a surface by dipping into a molten bath of zinc or Zn:Al alloy.

Hot filament CVD (HFCVD)   Chemical vapor deposition where a hot filament is used to decompose the precursor vapor. Used mainly to deposit diamond and diamond-like-carbon.

Hot isostatic pressure (HIP) (sintering)   Pressing of an object uniformly from all directions, usually in a hydrostatic media, at a high temperature. Used to form dense structures from powders.

Hot stamping   Application of a film pattern from a metallized web to a surface using pressure on a hot tool in the form of the pattern to be applied. The metallized web may have a release agent between the film and the web to aid in the release of the film on the die to the surface being stamped.

Hot-wall reactor (CVD)   Furnace where the CVD gases and the substrate are heated by conduction and radiation from the containing structure.

Hot water seal (anodization)   The hydration of anodized aluminum to cause the oxide to swell and seal the pores.

Housekeeping (contamination control)   Efforts to minimize contamination in the processing area. Examples include cleaning of surfaces, reducing clutter, storage in closed cabinets, no dust-catching surfaces such as the tops of cabinets, no spaces under cabinets that are hard to clean, etc.

Hume-Rothery's Rules of Solid Solubility (metallurgy)   They are: 1) complete miscibility can occur only if the unit cells of the two components are essentially alike, 2) if the diameters of the solute and solvent atoms differ appreciably, the solubility will be limited, 3) the closer the elements are to each other in the periodic table, the more likely they are to form a solid solution—conversely, the farther apart the elements are, the more likely they are to form a compound.

Humidity   The amount of water vapor in the air. See Dew point.

Humidity, absolute   The amount of water vapor in the air as measured in grams per cubic centimeter.

Humidity, relative   The ratio of the amount of water vapor in a gas to the amount it would hold at saturation expressed in percent.

Humidity shift (optical)   The change in optical properties of a material as a function of the humidity of the ambient environment.

Hybrid deposition system (PVD technology)   System using two or more deposition techniques in sequence, usually in separate chambers. See Deposition system.

Hybrid deposition process (PVD technology)   Deposition process that uses more than one deposition technique at the same time. Example: Reactive deposition of a carbonitride by sputtering a metal in a gas containing nitrogen, argon, and acetylene where the acetylene is decomposed in the plasma (VLP-PECVD) to provide the carbon, thus making a hybrid PVD/PECVD process.

Hybrid vacuum pump (vacuum technology)   Vacuum pump that combines more than one pumping mechanism. Example: A turbomolecular pump that has a molecular drag stage.

Hydration   Reaction of water such that the water molecules become an integral part of the chemical structure. Example: Anhydrous copper sulfate has the chemical formula $CuSO_4$, but the hydrated copper sulfate has the chemical formula $CuSO_4, 5H_2O$. See Anhydrous.

Hydrocarbon   Material composed of hydrogen and carbon bonded with the C=H chemical bond. Strict definition: only hydrogen and carbon. Loose definition: other atoms in molecule.

Hydrochlorofluorocarbon (HCFC) solvents (cleaning)   Solvent containing hydrogen as well as chlorine and fluorine. Examples: **HCFC-22** ($CHClF_2$); **HCFC-124** ($CHClFCF_3$). See Chlorinated solvents; Chlorofluorocarbon (CFC) solvents.

Hydrogen bridge bonding (chemical bonding)   When a hydrogen atom that is covalently bonded to one atom is bonded by polarization to another atom or molecule. Example: Water molecules bonded together to give liquid water.

Hydrogen embrittlement (cleaning)   When hydrogen is incorporated into the metal surface, making it more easily fractured. Hydrogen can come from acid cleaning of the surface.

Hydrogen plasma cleaning (cleaning)   Using a hydrogen plasma to promote reduction reactions or to hydrogenate hydrocarbons, thus making them more volatile.

Hydrogen reduction (cleaning)   The reaction of hydrogen with a material so as to give up an electron, often resulting in the decomposition of a molecule. Example: Hydrogen reduction of a metallic oxide to the metal, releasing water.

Hydrogenate (chemical structure)   Add hydrogen to a molecule.

Hydrophilic surface (cleaning)   Water-loving surface. Water will wet the surface.

Hydrophobic surface (cleaning)   Water-hating surface. Water will ball up and not wet the surface.

Hydrosonic cleaning (cleaning)   Hydrosonic cleaning utilizes hydrodynamically generated pressure waves to create agitation in the fluid-solid interface.

Hydrostatic weighing   Weighing in and out of a fluid of known density. This weight along with the measured volume allows determination of the density of the material.

Hydroxyl (radical)   The OH⁻ radical.

Hysteresis   The lagging of an effect behind its cause.

I

Ideal gas   A gas that is composed of atoms or molecules that physically collide but otherwise do not interact. Low-pressure gases are generally treated as ideal gases. Also called a **Perfect gas**. See Non-ideal gas.

Ideal Gas Law (vacuum technology)   An equation in gas kinetics that relates the volume (V), pressure (P), and absolute temperature (T) of an ideal gas ($PV = $ constant $\times$ T).

Imine   Class of compounds that have the NH radical (**Imine group**) attached to a carbon atom with a double bond.

Immersion cleaning   To leave the part in a cleaning solution for a period of time, often with mechanical movement of the part and/or agitation of the solution. Also called **Soak cleaning.**

Immersion heaters (cleaning, electroplating)   Electric heaters designed to be immersed into tanks of fluids to raise the temperature of the fluid. They may be bottom-mounted or side-mounted and may be coated to reduce corrosion. Typically 20-50 watts per square inch of heater surface.

Immersion plating   When an ion in solution that has a less negative potential than the atom of a solid in the solution spontaneously displaces the atom of the solid and deposits on the solid. Examples: Gold (+1.50 volts) plating onto copper (+0.52 volts); Lead (-0.126 volts) or tin (-0.136 volts) (from solder) plating on aluminum (-1.67 volts). Also called **Displacement plating**. See Electrochemical series.

Immiscible fluids   Non-soluble fluids.

Impact plating   Coating a surface by transfer of material from impacting particles on the surface. The par-

ticles may be at a high velocity or be pounded on the surface by a tumbling action. Also called **Mechanical plating**.

Impedance (electrical)   The resistance to flow of a current due to the ohmic resistance and the effects of inductance in the circuit.

Impedance matching, rf (plasma)   Matching the impedance of the load (plasma and electrode) to the impedance of the power supply in order to increase the power dissipated into the gas and minimize the power reflected back into the power supply.

Impregnation, vacuum   The removal of gases from pores in a material under vacuum, followed by coating the material with a fluid, then letting atmospheric pressure force the fluid into the pores.

Impurities (characterization)   Foreign materials that are present in a material. The impurities may or may not be detrimental or useful. See Dopant.

Impurities, major   Impurities in the amount of tenths of a percent or more.

Impurities, minor   Impurities in the amount of parts per thousand to parts per tenth.

Impurities, trace   Impurities in the amount of parts per thousand or less, down to parts per billion.

***In situ* cleaning** (PVD technology)   Cleaning in the deposition system. Examples: Ion scrubbing; Reactive plasma cleaning; Sputter cleaning.

In-chamber contamination (cleaning)   Contamination that occurs in the deposition system during pumpdown and vacuum processing. Example: Backstreaming of pump oils into the deposition chamber.

In-house coater (surface engineering)   A manufacturing facility that only coats items for one group that controls its actions. See Contract coater and Jobshop.

In-line processing system   An integrated processing system that uses several processing chambers connected together to sequentially process the substrates. The in-line systems are typically characterized by having the substrates moving from chamber-to-chamber in one direction so that substrates can be processed in each module all the time. See Cluster tool.

In-line processing system, valve-isolation   In the valve isolation in-line system there is a valve between processing chambers. In operation this valve has a very low pressure differential across the valve.

In-line processing system, pump-isolation   In the pump isolation in-line system there is an intermediate chamber ("tunnel") between the processing chambers. This intermediate chamber has a low conductance for gas flow between chambers and the region is actively pumped to prevent gases from one chamber getting into the other chamber.

In-line processing system, controlled-atmosphere transfer In the inert transfer in-line system the transfer chamber is at atmospheric pressure, so hermetically sealed gloves can be used. The gas in the transfer chamber can be a dry air if the product is moisture-sensitive, or an inert gas such as argon or nitrogen if chemical reaction is a problem.

In-line processing system, vacuum transfer In the vacuum transfer in-line system the fixture is moved into and out of a common transfer chamber that is under "rough" vacuum.

Inclusions (solids) Particles of second phase material found in a solid matrix. See Stringers.

Index of refraction The phase velocity of radiation in a vacuum divided by the phase velocity in a specific medium, usually vacuum. Examples: High index of refraction materials include TiO_2 and ZrO_2; Low index of refraction materials include air, SiO_2 and MgF_2. Also called **Refractive index**. See Refraction.

Induction heating Heating of an electrical conductor by placing it in a rapidly changing electric field so that the electrical currents are induced in the metal producing **Joule (I^2R) heating**.

Inductively coupled plasma (ICP) source (plasma) A plasma source where the plasma is formed in a region surrounded by an rf coil that couples energy into the electrons in the plasma.

Inert gas A gas that does not chemically react with surfaces under processing conditions. There include: "noble" gases that have filled electron shells (e.g., He, Ne, Ar, Kr, Xe) and thus are chemically inert, and other gases such as nitrogen under specific conditions.

Infrared (IR) spectrum Electromagnetic radiation in the wavelength range of 0.78 to 300 microns.

Infrared pyrometry Determination of the temperature of a surface by measuring the infrared radiation emitted from the surface. Useful in temperature ranges below where optical pyrometry (color temperature) is used. See Optical pyrometry.

Infrared window Material that has a high transparency for infrared radiation over some portion of the infrared spectrum. Examples: Sodium chloride; Silicon; Germanium; Potassium bromide (KBr); Cesium iodide; High-density polyethylene.

Inhibitor A chemical used to reduce the rate of a chemical or electrochemical reaction. Example: Rust inhibitor.

Inlet pressure (vacuum technology) The pressure at the inlet port of a vacuum pump.

Inspection, final (manufacturing) The final inspection before the completed device leaves the production area to ensure that it meets specified requirements. Also called an **Acceptance inspection**. See Process flow diagram.

Inspection, incoming (manufacturing) Inspection of the as-received material to ensure that it meets specifications before it enters the processing sequence. See Process flow diagram.

Inspection, in-process (manufacturing) Inspections at various stages of production to ensure that an unacceptable product is not being processed. The information can provide feedback into production processing before too much unacceptable material has been processed.

Installed cost (equipment) Cost to purchase and install the equipment. See Cost of ownership.

Insulated glass A glass structure that increases the R-value by having two or more panes of glass, separated by a small spacing to reduce convective heat transfer, and joined together at the edges, usually with a low-thermal-conductive structure that often contains a desiccant. Often the glass is coated with a low-E film.

Insulator (electrical) A material with a low electrical conductivity and the ability to prevent arcing (voltage breakdown) between materials at different electrical potentials as measured in volts per mil.

Interface (film formation) The region of contact between two materials. See Interphase material.

Interface, abrupt The interface that is formed between two materials (A and B) when there is no diffusion or chemical compound formation in the interfacial region. The transition of A to B in the length of a lattice parameter ($\approx$3Å).

Interface, combination An interface composed of several types of materials such as an alloy with a second phase dispersed in it.

Interface, compound When the interfacial material (interphase material) that has been formed during the deposition of A onto B, along with subsequent diffusion and reaction, consists of a compound of A and B such as an intermetallic compound.

Interface, diffusion When the interfacial material (interphase material) that has been formed during the deposition of A onto B, along with subsequent diffusion, consists of an alloy of A and B with a gradation in composition. See Kirkendall porosity; Interphase material.

Interface, mechanical interlocking A "tongue-and-groove" interlocking where the materials "key" into each other at the interface and a fracture that follows the interface must take a circuitous route with greatly changing stress tensors as the fracture propagates.

Interface, pseudodiffusion An interfacial region where the material is graded, similar to the diffusion interface. Produced by mechanical means such as beginning the second deposition before stopping the first deposition, or by implantation of high energy "film ions."

Interfacial flaws (film formation, adhesion)   Flaws, such as microcracks or voids, that reduce the fracture strength of the interphase material.

Interference, constructive   When radiation from two sources interact with each other such that the amplitudes add together to produce an intense signal. Example: The white band in optical interference patterns.

Interference, destructive   When radiation from two sources interact with each other such that the amplitudes subtract to produce a weak signal. Example: The dark band in optical interference patterns.

Interferometer   An instrument that measures interference effects using either monochromatic radiation and/or white (continuum) radiation.

Interlock (vacuum technology)   A device that prevents a component from operating normally if it does or does not get a signal from a sensor indicating that something is not correct. Example: Electrical interlock that prevents a high voltage from being applied to a sputtering cathode unless the system is under vacuum as indicated by a pressure sensor.

Intermetallic compound   A chemical compound composed of two metals, one of which is an amphoteric material. Example: Al_2Cu where aluminum is the amphoteric material. See Amphoteric material.

Interphase material (adhesion, film formation)   The material at the interface that is formed by diffusion, reaction, or co-deposition at the interface between the film and the substrate. The properties of this material are an important consideration in adhesion. Also called **Interfacial material**.

Interstitial (crystallography)   A position between normal lattice sites. Example: An interstitial atom of carbon dissolved in a metal lattice.

Intertool transport   Movement between one tooling arrangement and another tooling arrangement. Often between chambers separated by an isolation valve. See Tooling.

Ion   An atom or molecule that has an excess (**Negative ion**) or deficiency (**Positive ion**) of electrons.

Ion assisted deposition (IAD) (film deposition)   Concurrent or periodic bombardment with energetic reactive ions during film deposition. See Ion plating. When using an ion beam the process is sometimes called Ion beam assisted deposition.

Ion beam assisted deposition (IBAD) (film deposition)   A special case of ion plating where the deposition is done in a high vacuum and the concurrent or periodic bombardment is provided by gaseous ions accelerated from an ion gun or plasma source. Also called **Vacuum-based ion plating; Ion beam enhanced deposition; Ion assisted deposition (IAD).**

Ion beam deposition (film deposition)   Deposition of a film using ions of the film material, usually obtained by the decomposition of a vapor precursor in a plasma source. Example: Deposition of i-C from methane decomposed in a plasma source.

Ion beam enhanced deposition (IBED)   A special case of ion plating where the deposition is done in a high vacuum and the concurrent or periodic bombardment is provided by ions accelerated from an ion gun or a plasma source. Also called **Ion beam assisted deposition (IBAD)** (preferred).

Ion beam mixing (adhesion)   The mixing across an interface to increase film adhesion by high-energy ions that penetrate through the interfacial region. Also called **Interfacial stitching**.

Ion beam sputtering   Physical sputtering using an energetic ion beam from an ion gun in a good vacuum.

Ion cluster beam (ICB) deposition (PVD technology)   A deposition process in which clusters of atoms (1000s of atoms) are electrically charged and accelerated to the substrate to deposit with greater than thermal energy.

Ion exchange (water purification)   The exchanging of Na^+ or H^+ ions for positive ions and Cl^- or OH^- ions for negative ions in hard water to produce soft (Na^+, Cl^-) or ultrapure (H^+, OH^-) water. See Reverse osmosis.

Ion implantation   The physical injection of high energy (MeV) ions into the surface region of a material to change the electrical (**Doping**) or mechanical properties of the near-surface region.

Ion milling   The machining (removal) of material by sputtering.

Ion plating (PVD technology)   There is no universally accepted definition of the term "ion plating." Ion plating can be defined as a film deposition process in which the growing film is subjected to concurrent or periodic high-energy ion bombardment in order to modify film growth and the properties of the deposited film. The term does not specify the source of depositing atoms (sputtering, thermal evaporation, arc vaporization, chemical vapor precursors, etc.) nor the source of bombarding species (plasma, ion gun, plasma source, etc.) or whether the bombarding species is reactive, non-reactive, or a "film ion." Other definitions restrict the configuration to using an evaporation source or a DC diode plasma. Also called **Ion assisted deposition (IAD); Ion vapor deposition (IVD).**

Ion plating, arc   Ion plating where the source of vaporized material is from arc vaporization.

Ion plating, chemical   Ion plating where the source of depositing material is from a chemical vapor precursor species such as CH_4.

Ion plating, reactive   Ion plating in a reactive gaseous environment where a film of a compound material is deposited.

Ion plating, sputter (SIP) Ion plating where the source of vaporized material is from sputtering of a solid surface.

Ion plating, vacuum See Ion beam assisted deposition (IBAD).

Ion polishing Polishing a surface by high-angle sputtering of a rotating surface.

Ion pump (vacuum technology) A high-vacuum pump that operates by sputtering a reactive getter material such as titanium, which then reacts with the reactive gases in the system. Inert gases are pumped by being implanted and buried in the depositing material. See Vacuum pump.

Ion scattering spectrometry (ISS) (characterization) A surface analytical technique in which the probing species are energetic ion species with a specific energy and the detected species are reflected ions that have lost specific amounts of energy by collision with the surface atoms.

Ion scrubbing (cleaning) The desorption of adsorbed species from a surface in contact with a plasma under the action of ions accelerated across the plasma sheath.

Ion source (plasma technology) A device for generating ions. Often an ion beam is formed by extraction of ions, using a grid system, from a plasma source and the ions are accelerated away from the source. See Plasma source.

Ion vapor deposition (IVD) Ion plating generally using aluminum as the film material. Terminology used mostly in the aerospace industry. **See Ion plating.**

Ionic bonding Chemical bonding between electrically charged ions.

Ionitriding (surface modification) The bombarding of a hot surface with nitrogen ions in order to inject the nitrogen into the surface and enhance diffusion into the surface to form a hard case. See Gas conversion; Plasma immersion ion implantation (PIII).

Ionization The formation of ions, generally by electron-atom/molecule impact. Other processes, such as Penning ionization, can also cause ionization.

Ionization deposition rate monitor (PVD technology) A deposition rate monitor that compares the collected ionization current in a reference ionizing chamber to the collected ion current from an ionizing chamber through which the vapor flux of the film material is passing.

Ionization gauge (vacuum technology) A vacuum gauge that uses ion current formed by electron-atom collisions as an indicator of the gas pressure (density). The electrons are formed as secondary electrons from ion bombardment or from a hot thermoelectron emitting filament. See Vacuum gauge.

Isentropic process A process without a change in entropy. See Entropy.

Island-channel-continuous (film formation) The development of a continuous film under Volmer-Weber nucleation conditions where isolated nuclei grow in size, contact each other, then fill in to form a continuous film.

Isobaric process A process without change in pressure.

Isolation technology (contamination control) A set of technologies and procedures that isolate a product from ambient contamination during processing and transportation.

Isomer Variation of atomic arrangement (and properties) in molecules having the same atomic composition. Example: Normal propyl alcohol and isopropyl alcohol C_3H_8O.

Isothermal process A process without change in temperature.

Isotropic property (characterization) A property that is equal in all directions. See Anisotropic property.

Issue (document) A dated or sequential version of a document such as a specification.

J

Jet assembly (diffusion pump) The arrangement of surfaces in a diffusion pump that imparts a preferential direction to the vapors formed by heating the pump fluid. Also called a **Nozzle assembly**.

Jet vapor deposition (film deposition) An atomistic deposition process where evaporated atoms are introduced into a supersonic jet flow of inert carrier gas that transports the atoms to the substrate surface.

Jobshop (surface engineering) A manufacturing facility that will coat items for any individual requiring its services. Also called **Contract coater**. See In-house coater.

Joule (J) The SI unit of work, energy, heat impulse, and momentum.

Joule heating Resistive heating given by I^2R where I is the electrical current and R is the resistance of the conductor.

К

Karat A unit for defining the purity of gold, with 24 karat being pure gold.

Kaufman ion source (plasma) An ion source that uses a grid system to extract ions from a confined plasma established using a thermoelectron-emitting filament in a magnetic field.

Kelrez™ (vacuum technology) An elastomer that is

more chemically stable than Viton™. Used in plasma etching systems.

Kelvin (K) temperature scale   A temperature scale defined as zero degrees Kelvin, being the temperature at which there is no molecular motion and the heat content of the material is zero. The Kelvin degree has the same magnitude as the Centigrade degree. The triple point of water is then 273.16 K. Zero degrees K = -273.16°C and -459.67°F.

Keyholing (metallization, semiconductor)   When the opening of a high-aspect-ratio hole or trench closes during film deposition before the bottom of the hole or trench is filled. See Mouse hole.

KF flange (vacuum technology)   An O-ring sealing flange with a specific clamping configuration. See MF flange.

KiloGray (kGy)   The amount of nuclear radiation that will raise the temperature of the object 0.43°F.

Kinetic energy   Energy due to motion. See Potential energy.

Kirkendall porosity (film formation, adhesion)   Porosity that develops in the interfacial region between two materials when the first material diffuses faster into the second than the second diffuses into the first, thus producing a loss of mass and formation of voids in the interfacial region. Also called **Kirkendall voids**.

Knife-edge (vacuum technology)   The sharp edge used to shear into a soft metal gasket to provide vacuum sealing. Example: Knife-edge on a CF flange that cuts into a soft copper gasket.

Knob-twiddler (manufacturing)   A person who has a propensity for changing things, often in disregard of the Manufacturing Process Instructions (MPIs).

Knock-down filter (vacuum technology)   Surface used to reduce the velocity of high-velocity particles in the exhaust side of an etching or CVD system. Also called a **Knock-down plate.**

Knoop (HK) hardness number   The expression derived from the force used and the projected area of an imprint obtained by a specifically shaped (ASTM E 384) diamond indenter forced into a surface. Abbreviated HK (formally KHN). HK = 14,229 P/d^2 where P = grams force and d = length of long diagonal in microns. See Vickers hardness number.

Knudsen cell (PVD technology)   A thermal vaporization source that emits vapor through an orifice from a cavity where the vapor pressure is carefully controlled by controlling the temperature. Used in Molecular beam epitaxy (MBE) processing. Also called an **Effusion cell.**

Knudsen flow (vacuum technology)   The transition gas flow range between viscous flow and molecular flow.

Knurling   Impressing a design into a surface by deformation using a roller with a hardened surface containing a design in relief. The process results in workhardening the surface. See Coining.

Kosher electroplating   Electroplating using kosher additives.

Krypton 85 (leak detection)   Radioactive (12 keV beta emitter) isotope of the element krypton (amu = 84). Used for high-precision leak detection (basic sensitivity of about 10^{-13} std cm^3/s). See Helium.

L

Labile structure (crystallography)   A crystallographic structure that is readily changed by heat or some other process. Also called a **Metastable structure.**

Lacquer (topcoats)   A solution of organics in a solvent that crosslinks and forms a film when the solvent is evaporated. In the early 1900s terminology, the solid material was nitrocellulose-based and the solvent was non-aqueous. In newer uses of the term, the solid can also be a thermoplastic material, such as a vinyl or an acrylic, or thermosetting materials, such as epoxies and phenolics, and the solvent can be water.

Lacquer coating (decorative coating)   The topcoat that is used to give abrasion and corrosion resistance, color, and texture to a decorative coating system. The lacquer is typically applied over a reflective aluminum film deposited by vacuum evaporation that may be deposited on a flow-coated basecoat, which creates a smooth surface. See Basecoat; Topcoat.

Laminar flow (cleaning)   Gaseous flow in the viscous flow range but with little turbulent mixing.

Laminated glass (automotive, architectural)   A sandwich of two panes of glass with a polymer interlayer that is used to prevent the glass from fracturing into jagged pieces when broken and to give structural integrity even when broken. Also gives better sound attenuation than solid glass. Interlayer may contain wire (**Wired glass**) for further structural integrity.

Lamination   The bonding of two or more layers together, usually by heat and pressure or by using an adhesive.

Langmuir probe (plasma)   A small-area non-disrupting probe that is used to measure the electron density and the electron temperature in a plasma.

Larmor radius (plasma)   The radius of the path that an electron takes in a magnetic field. Also called the **Gyro-radius**.

Laser   Term used synonymously with the acronym for light-amplification by stimulated emission of radiation (LASER).

Laser ablation (vaporization)   Vaporization by the adsorption of energy from a laser pulse. Also called **Laser vaporization**.

Laser ablation deposition (LAD) (film deposition) PVD using laser vaporization as the vapor source. Also called **Pulsed laser deposition (PLD).**

Laser cleaning Removal of contaminants from a surface using a laser to provide thermal energy by photoadsorption to desorb the contaminant or to vaporize some of the surface or to volatilize water trapped between the particle and the surface.

Laser enhanced CVD Increasing the reaction rate using a laser to provide thermal energy by the adsorption of radiation by the substrate or by **Photodecomposition** of the chemical vapor precursor.

Laser glazing A method of rapidly melting and cooling a surface or a film on a surface. Used to densify and smooth the surface and to enhance interdiffusion and reaction.

Laser melt-particle injection Process where the surface is melted with a laser and metal carbide particles are mixed with the molten pool before solidification.

Laser treatment (glazing, annealing, crystal structure modification) A method of rapidly heating and cooling a surface in order to densify a surface, refine the grain size, crystallize an amorphous material, etc. Example: Laser treatment of an amorphous silicon film to convert it into a polysilicon material.

Latex (cleaning) Often used synonymously with rubber. Example: Latex (rubber) gloves.

Lattice, crystal (microstructure, crystallography) The regular, periodic arrangement of atoms in a crystalline solid. See Crystal structure.

Lattice defects (crystallography) Discontinuities in the lattice structure such as **Vacancies, Interstitial atoms, Substitutional atoms,** and **Dislocations.**

Lattice misfit (film formation) When the lattice of the substrate does not have the same spacing as the film material being deposited. Small misfits can be accommodated by lattice strain (**Strained-layer superlattice**). Large misfits cause dislocations in the interfacial region that extend through the film.

Lattice parameter (crystallography) The atomic separation in a crystalline solid. See d-spacing.

Leaching Preferential chemical removal of one constituent to produce surface depletion of that material and surface enrichment by the remaining material. The resulting surface may be porous or, in the case of metals, may be burnished to densify the surface.

Leak, real (vacuum technology) A conduction path from the external ambient environment into a vacuum system.

Leak, virtual (vacuum technology) A conduction path from an internal trapped volume to the main volume of a vacuum system (no connection to the outside ambient environment). Example: Void below a solid bolt in a blind, tapped hole.

Leak detection (vacuum technology) The process of finding a leak in a vacuum system. See Helium leak detector.

Leak rate (vacuum technology) The amount of gas passing through a leak expressed in Torr-liters/sec.

Leak valve (vacuum technology) A device used to introduce gas into a system in a controlled manner. See Valve, vacuum.

Leak-tight system (vacuum technology) A vacuum system that has a leak rate less than a specified value using a specific leak-detection gas and defined leak-detection techniques.

Leak-up rate (vacuum technology) The time for the pressure in a system to rise a specified amount with no vacuum pumping taking place. Generally the leak-up pressure range is specified, i.e., from 10^{-4} Torr to 10^{-3} Torr. The leak-up rate is an indication of the presence of outgassing, desorption, virtual leaks and real leaks. Also called **Rate-of-rise.**

Legs (cleaning) The flow of a fluid as it avoids contaminated areas on a surface to give thick, often narrow, flow streams.

Lehr (glass) An annealing furnace for glass.

Lever rule (phase diagram) A method of determining the relative compositions of an alloy under stable conditions at a specific temperature.

Lewis acid A material that acts as an electron acceptor.

Lewis base A material that acts as an electron donor.

Life-test (characterization) Evaluation of a function or property under specific conditions that simulate service conditions, in order to determine how long it will function correctly. See Shelf life.

Life-test, accelerated (characterization) Evaluation of a property or a function under conditions that will accelerate failure and allow the determination of the activation energy for failure. By using the Arrhenius relationship, the failure time under less severe conditions can be calculated provided the activation energy for failure and the failure mode remain constant. See Arrhenius equation.

Lift-off (patterning) Forming a pattern by first depositing a pattern of material to be removed such as a photoresist, then depositing a blanket film of the desired film material, then removing the first pattern, leaving the desired pattern.

Limiting foreline pressure (vacuum technology) The outlet pressure of a pump above which the pumping efficiency of the pump rapidly deteriorates. See Crossover pressure.

Liner, chamber (PVD technology) A removable surface in a chamber used to collect vaporized material and prevent it from depositing on non-removable surfaces.

See Non-removable surfaces; Vacuum surfaces.

Liner, pocket (e-beam evaporation) A crucible-like container that is sometimes used in the pocket of the e-beam evaporation hearth to lower the conductive heat-loss from the melt and to allow easy removal of the charge from the hearth.

Lint (cleaning) Small particles of organic material usually formed by breaking off the ends of fibers.

Liquefication by compression (vacuum technology) When compression results in the partial pressure of a vapor exceeding the saturation vapor pressure, producing condensation of the excess vapor into a liquid. Example: Water vapor compressed to a pressure above 20 Torr at room temperature will liquefy the excess vapor until the pressure becomes 20 Torr.

Liquid honing Producing a polished surface by abrasion using fine abrasive particles entrained in a high-velocity liquid stream.

Liquid jet pump (vacuum technology) A kinetic vacuum pump where the gases are entrained in a stream of fluid or steam. See Steam jet pump; Verneuli tube.

Liquid-like behavior, nuclei (film formation) The ability of nuclei to move and rotate on a substrate surface.

Liquidus (phase diagram) The upper boundary of the liquid + solid region of the phase diagram. See Solidus.

Liquidus range The temperature range between the melting point and the boiling point of a material.

Load, pumping (vacuum technology) The amount of gas (mass flow) passing through the vacuum pump.

Load-lock system (processing) A processing system where the substrates are introduced ("loaded") into a chamber where some processing is done such as heating, rough-pumping, or outgassing, then the materials are moved through a valve into a separate chamber for further treatment, such as film deposition.

Loading factor (PVD processing) A processing variable that is the dependence of the processing parameters on the number of substrates, or the total surface area of the substrates being processed.

Log, calibration (manufacturing) A dated record of who, when, and how calibration was performed on a piece of equipment.

Log, maintenance (manufacturing) A dated record of when and what maintenance was performed on a piece of equipment and who performed the maintenance.

Log, operation (manufacturing) A dated record of when a system was used. This, together with the maintenance log, allows establishing the time between routine cleaning and maintenance operations. Also called a **Run log**.

Long-focus electron beam (evaporation) A high-power electron gun that allows heating and evaporation by focusing an electron beam on the surface from a source that is a long distance away and without bending the electron beam. Example: Pierce gun. See Deflected electron beam.

Lot (PVD technology) All of the materials (substrates, source material, etc.) of identical purity, structure, composition, etc., obtained in a single shipment and traceable to a specific manufacturer.

Low-carbon steel (vacuum technology) A low-cost, ductile, non-hardenable iron alloy that contains a low concentration of carbon. Often used for large vacuum chambers. Care must be taken to avoid corrosion (rust) with use.

Low-pressure CVD (LPCVD) (vacuum deposition processes) Chemical vapor deposition that is performed in a vacuum. Also called **Sub-atmospheric CVD**. See Chemical Vapor Deposition; Plasma Enhanced CVD.

Low pressure PECVD (LPPECVD) (vacuum deposition processes) Plasma enhanced CVD performed at a low enough pressure (10-20 mTorr) that high-energy ion bombardment effects can be important. See Plasma Enhanced CVD.

Low-pressure plasma spraying (LPPS) Plasma spraying that is performed in a vacuum. See Plasma spraying.

Low-E coating (glass) A low-emissivity thin film structure that is used on glass or a polymer film in a double-pane window to reflect infrared energy back into a room. A typical thin film structure might be: ZnO-Ag-(Ti)-ZnO-TiO_2 where a very thin film of titanium (Ti) is used to protect the silver layer (120Å) during the reactive deposition of ZnO, which decreases the reflectance of the silver in the visible. The TiO_2 is an abrasion-resistant topcoat. See Solar control coating.

Low-K film (semiconductor processing) A low-dielectric-loss film.

Low-U (glass) Low heat transfer glass.

Lubricant (vacuum technology) A lubricating liquid or solid material that is vacuum-compatible. Examples: MoS_2 dry lubricant; Silicone greases. Note: Graphite is not vacuum-compatible as a lubricant.

Lusec (leak detection, *Archaic*) A leak rate of 1.0 liter-micron-sec[-1] (l-μ-sec[-1]), where $\mu = 10^{-3}$ Torr.

M

M classification (contamination control) Classification of a cleanroom as to the number of particles per cubic meter that have a size greater than 0.5 micron. Expressed as the logarithm of the number to the base 10. See Class.

Machine direction (web coating) Direction that the web is moving. See Transverse direction.

Macrocolumnar morphology (film formation) The

large-sized columnar morphology that develops due to the initial surface features of the substrate. See Columnar morphology.

Macros (arc vaporization) Molten globules of electrode material ejected under arcing conditions from a solid cathode and deposited onto the substrate, giving nodules in the film. See Filtered arc source; Plasma duct.

Magnetic induction (film thickness) Used to measure the thickness of a non-magnetic film on a ferromagnetic surface. The probe measures the magnetic coupling between an AC-driven primary through the substrate to the secondary of a transformer configuration. The lift-off caused by the film changes the magnetic coupling.

Magnetoresistance The change in electrical resistance as a function of the magnetic field.

Magnetron A crossed-field electromagnetic system where the path of electrons accelerated in an electric field is controlled by a magnetic field at an angle to the electric field. In a magnetron tube the electron motion is used to generate microwave radiation (**Klystron tube**). See Magnetron.

Magnetron (sputtering) Sputtering using a crossed-field electromagnetic configuration to keep the ejected secondary electrons near the cathode (target) surface and in a closed path on the surface. This allows a dense plasma to be established near the surface so that the ions that are accelerated from the plasma do not sustain energy loss by collision before they bombard the sputtering target. The closed path can be easily generated on a planar surface or on any surface of revolution. Also called a **Surface magnetron**.

Magnetron, conical A magnetron configuration where the target surface is the interior surface of a truncated conical section. The anode is often positioned in the region of the small-diameter portion of a doubly truncated cone. Also called an **S-gun**.

Magnetron, hemispherical A magnetron configuration where the target surface is the interior surface of a hemispherical section. The anode is often positioned around the lip of the hemisphere.

Magnetron, hollow cathode A magnetron configuration where the target surface is the interior surface of a cup. Permanent and moving magnets are used to shape the magnetic field in the cup.

Magnetron, hollow cylinder A magnetron configuration where the target surface is the interior surface of a hollow cylinder. The cylinder often has a flange at each end to prevent loss of electrons.

Magnetron, planar A magnetron configuration where the target surface is a planar surface and the magnetic field is in a configuration such that it is round or oval. The sputter-erosion track resembles a "**Racetrack**."

Magnetron, post A magnetron configuration that is a post, perhaps with flares on the ends (spool), with a magnetic field either axial to the post or in a series of looped magnetic fields around the post. The electrons are confined along the surface of the post and between the flared ends. Also called a **Spool magnetron**.

Magnetron, rotatable cylindrical A planar-like magnetron configuration where the target surface is the exterior surface of a hollow water-cooled tube that is rotated through the magnetic field.

Magnetron, unbalanced (sputtering) A magnetron configuration in which the magnetic fields are arranged so as to allow some of the secondary electrons to escape from the vicinity of the cathode in order to establish a plasma between the target and the substrate.

Magnetrons, dual AC Two planar magnetrons that are side-by-side and are alternately the cathode and the anode of an AC (>10 kHz) voltage. This arrangement eliminates the **Disappearing anode** effect in reactive sputter deposition. See Mid-frequency.

Magnetrons, dual unbalanced Two unbalanced planar magnetrons positioned such that they face each other with the surface to be coated positioned between the two magnetrons. Generally the north escaping field of one magnetron faces the south escaping field of the other magnetron.

Makeup (water, chemicals) (cleaning) Water or chemicals needed to bring the material in a cleaning tank back to some previous level or concentration. Example: Water can be lost by **Carryover** or evaporation from a rinse tank.

Mandrel (electroplating, CVD, PVD technology) A form (substrate) on which a coating is deposited that is subsequently removed, leaving a free-standing structure. See Vapor forming.

Manometer, liquid (vacuum technology) A pressure-measuring device that uses a liquid column to measure the pressure difference in two volumes of gas. Often "U" shaped (two-legged) with a good vacuum above one of the legs and the gas being measured above the other leg.

Manufacturability The issues involved in commercially producing an item, including patent position, availability of raw materials, availability of components from outside suppliers, availability of suitable manufacturing space, scale-up, costs, etc. See Scale-up.

Manufacturing, early Manufacturing in the early stages where there are numerous experiments to fine-tune the processing parameters and equipment development to improve product yield and throughput. This can result in many changes to the **Process documentation**.

Manufacturing, mature Manufacturing after the equipment and processes have been optimized and there are few changes to the **Process documentation**.

Manufacturing process instruction (MPI) Detailed instructions for the performance of each operation and the use of specific equipment, based on the specifications, that apply to each stage of the process flow. MPIs are developed based on the specifications. See Process flow diagram; Specifications

Marangoni Principle The Marangoni Principle states that a flow will be induced in a liquid body where there are different surface tensions. For example, if a surface is wetted by water and is slowly withdrawn from water, a meniscus will form. If alcohol is present in the atmosphere above the water, the concentration of the alcohol will be greater in the meniscus than in the bulk of the water. This will create a difference in the surface tension of the water and the water/alcohol mixture will be pulled from the surface into the bulk of the water.

Mask (PVD technology) A physical cover that prevents film deposition on an area of the substrate surface. The mask may be in contact with the surface or in the line-of-sight from the source to the substrate. See Mask, moving.

Mask, moving (film formation) A method of forming a film structure having a specific thickness distribution by using a moving mask to determine the area and time on which the film material is being deposited on specific areas of the substrate.

Mass A measure of the resistance of a body to being accelerated. Term is often used synonymously with weight but that is not rigorously correct. See Weight.

Mass flow controller (MFC) (vacuum technology) A component that uses the output of a mass flow meter to control the conductance of a valve and thus control the gas flow through the gas manifold. The component is usually located upstream from the deposition chamber but can be located downstream from the chamber.

Mass flow meter (MFM) (vacuum technology) A component that measures the mass flow of a gas through a manifold system, usually by measuring the heat transfer. See Mass flow controller.

Mass spectrometer A device that determines the charge-to-mass (e/m) ratio of ionized species by deflecting them in an electric or magnetic field or by determining the "time-of-flight" between points in an accelerating electric field. See Partial pressure analyzer; Residual gas analyzer; Quadrupole mass spectrometer.

Mass spectrum The output of the mass spectrometer showing the position and height of the ion current resulting from the collected masses with a specific charge-to-mass (e/m) ratio.

Mass throughput (vacuum technology) The mass (grams per second) or number density (atoms or molecules per second) of gas that passes through a system or a component. Also called **Mass flow rate.**

Material safety data sheet (MSDS) (safety) A sheet available from the manufacturer for all chemicals used in the workplace that details the chemical composition, hazards, and potential hazards associated with using the material. By law the MSDSs must be made available to the workers exposed to the chemicals.

Material test report (MTR) (semiconductor processing equipment) A document that accompanies each lot of stainless steel tubing that provides the chemical composition, mechanical properties, etc., and is used to determine the welding parameters.

Maxwell velocity distribution The statistical velocity distribution of gas molecules at a given temperature showing the variation of higher and lower velocity (energy) particles from the average velocity. See Boltzmann's constant.

May Term used in a Specification or an MPI that grants permission. Example: The gloves may be reused. See Should; Shall.

Mean free path The average distance that a molecule travels between collisions with other molecules.

Mechanical activation (cleaning) Mechanical disruption of the surface barrier layers, such as oxides, to expose the underlying material and increase chemical reaction rates with the surface. Example: Brushing with a stiff metal wire brush in the deposition system just prior to film deposition.

Mechanical disruption (film growth) A means of disrupting the columnar growth mode by periodically deforming the surface mechanically, such as by burnishing.

Mechanical filter (contamination control) A filter that prevents the passage of particles by having very small passages in the filter media. Example: HEPA filter.

Mechanical interlocking-type interface (film growth, adhesion) A "tongue-and-groove" interlocking where the materials "key" into each other at the interface. A fracture that follows the interface must take a circuitous route with greatly changing stress tensors. See Interface.

Mechanical polishing Abrasive removal of the high points on a surface.

Mechanical pump (vacuum technology) A compression-type vacuum pump with moving parts. The term is generally applied to pumps used for roughing or backing (e.g., oil-sealed mechanical pump, piston pump, diaphragm pump, etc.) and not high-vacuum pumps (e.g., turbomolecular pumps). See Vacuum pump.

Mechanical scrubbing (cleaning) Rubbing a surface with a cloth or sponge, usually wet or under a liquid. The scrubbing action displaces contamination from the surface but care must be taken that the scrubbing action does not result in abrasive transfer. To avoid abrasive transfer the rubbing pressure should be controlled. See Abrasive transfer.

Mechanical working (fatigue)   The fatiguing of a metal by periodic mechanical deformation.

Mechanical working (forming)   The shaping of metals by deformation such as rolling, forging, or extrusion. (This type of processing generally creates a texture to the grain orientation.)

Medical air   Pure air with no oils or other contaminants that would affect the lungs of an individual breathing the air. Also called **SCBA** (self contained breathing apparatus) **air**.

Medium vacuum (vacuum technology)   The pressure range between rough vacuum and high vacuum.

Megasonic cleaning (cleaning)   Cleaning by high-frequency (> 400 kHz) pressure waves in a fluid where there is no cavitation. The cleaning action is due to frictional drag of the fluid moving over the surface. Used in cleaning flat surfaces such as wafers in semiconductor processing.

Meissner trap   A cryogenically cooled surface located in the processing chamber to condense water vapor without it having to enter the pumping manifold. Named after C.R. Meissner, who first reported the use of the technique in 1954. Also called a **Cryopanel**; **Cryocoil**. See Polycold.

Melt (material)   A specific lot of material made by melting. Example: Melt #— of stainless steel.

Melt (phase change)   Convert from a solid to a liquid.

Melt smoothing (surface modification)   Smoothing of a surface by melting since the molten surfaces tends to become smooth by surface tension effects.

Mer   The repeating structure unit in a polymer. See **Monomer**.

Mesh sizing   Obtaining particles with a specific size distribution by passing the particles through a series of screens having a specific number and size of openings per square inch. Particles that pass through one mesh, but not the next, have a specific size range.

Metalizing (decorative coating)   The least-preferred spelling of metallizing.

Metallic bonding   The chemical bonding resulting from metallic ions being immersed in a continuum of electrons. See Chemical bond.

Metalliding (electroplating)   Electroplating in a high-temperature molten salt bath where the deposited material diffuses into the surface of the part.

Metallization (decorative coating)   To apply a metal film, usually aluminum, to a low-cost part—often a molded plastic or zinc die-cast part. Also called **Junk coating**.

Metallization (electronics)   Application of an electrically conductive film to a non-conductive surface.

Metallization (general)   Application of a metal film to a surface.

Metamerism (optical)   Obtaining the same color from two different spectra.

Metastable state   A state that can easily be changed. Examples: Metastable excited state; Metastable crystallographic structure.

Meter (measurement)   A unit of length in the MKS system of measurement. Initially the meter was equal to one-ten-millionth of the distance of the Earth's quadrant on a meridian that passed through Paris, France (1793).

Methane (CH$_4$)   A gas that is used as a chemical vapor precursor for carbon in reactive deposition processes.

Metrology (surfaces)   The science of measurement.

MF flange (vacuum technology)   An O-ring sealing flange that uses a specific clamping configuration.

Mho   A unit of conductance equal to the reciprocal of the resistance in ohms. See Siemens.

Micelle (cleaning)   A cluster or aggregate of molecules. Example: Surfactant molecules agglomerating into micelles.

Micro-X-ray analysis (characterization)   When a collimator is used to define the area of analysis down to as small as 50 microns diameter for X-ray fluorescence analysis.

Microcolumnar morphology (film formation)   The morphology that develops with thickness due to the development of surface roughness due to preferential film deposition on high points on the surface. The columnar morphology resembles stacked posts and the columns are not single grains. Also called **Columnar morphology** (preferred). See Macrocolumnar morphology.

Micron (length)   Micrometer or 10^{-6} meter, 10^3 nanometers, 10^4 Ångstroms.

Micron (pressure)   Pressure unit equal to 10^{-3} Torr.

Microstructure (film)   The crystallography, grain size, phase distribution, lattice defect structure, voids, etc., of a film as determined by using an analytical technique such as transmission electron microscopy (TEM). See Morphology, film.

Microwave   There is no sharp distinction between microwave frequency and radio frequency (rf) waves or infrared radiation, but typically microwaves are in the 1 to 100 gigahertz (GHz) range with a wavelength shorter than about 30 centimeters. A common industrial microwave frequency is 2.45 GHz.

Mid-frequency (sputtering, ion plating)   The AC voltage frequency range of 10-250 kHz used for substrate biasing and sputtering. This frequency range can be generated using solid-state electronics.

Mil   One-thousandth of an inch or 25 microns.

Mill finish (metal)   The finish on a metal as it emerges from the fabrication mill. Example: Mill oxide scale.

Miller indices (crystallography) Nomenclature defining crystallographic planes in a crystal. Example: (111) plane in a cubic crystal.

Mirror A smooth surface that has spectral reflectivity and no distortion of an image on reflection.

Mirror-grade glass A glass that is flat enough to give no visual distortion of the reflected image when coated to make a mirror. The glass will also have no defects such as **Seeds** or **Stones**.

Miscible Capable of forming a stable uniform dispersion of one material in another to some **Solubility limit**. Also called **Soluble**.

Mixture (atomic) A uniform dispersion of two or more atomic species in one another. If the mixture is thermodynamically stable the mixture is called an **Alloy**.

Modified surface A surface that has properties different than the bulk, and the bulk material is detectable in the modified surface. Surface modification can be done chemically, electrochemically, mechanically, etc. Examples: Anodized aluminum; Shot-peened surface.

Modulus of elasticity The ratio of the applied tensile stress to the resulting elastic strain. Also called **Young's Modulus.**

Moisture transmission rate (MTR) The amount of moisture transmitted through a film in units of amount per unit area per unit time. Also called **Water vapor transmission rate (WVTR)** (preferred).

Molality (m) (chemistry) Concentration of a solution expressed in moles of solute per kilogram of solvent.

Molar solution (chemistry) A solution that contains one mole (gram-molecular weight) of the solute in one liter of the solvent.

Mold release (cleaning) A coating applied to a mold to minimize adherence between the mold surface and the molded part. The mold release is often a silicone and leaves a contaminant on the surface of the molded part that is very difficult to remove.

Mole (chemistry) The amount of a material whose mass in grams is equal to the molecular weight. A mole contains 6.023×10^{23} chemical units (atoms or molecules). Also called **Gram molecular weight**. See Avogadro's Law.

Mole fraction (chemistry) The number of moles of a substance in a material divided by the total number of moles of all substances in the material.

Molecular beam epitaxy (MBE) The epitaxial growth of a single-crystal film produced in a very good vacuum using a well-controlled beam of atomic or molecular species that is usually obtained by thermal evaporation from an effusion cell. See Knudsen cell.

Molecular drag pump (vacuum technology) A kinetic vacuum pump in which velocity is imparted to the gas molecule by contact with a high-velocity surface. See Vacuum pump.

Molecular flow (vacuum technology) Flow condition where there are few collisions between molecules because of the long mean free path for collision (low pressure).

Molecular sieve (vacuum technology) An adsorbent material characterized by a high surface area formed by having many small pores of a well-defined size. See Zeolites; Activated carbon.

Molecular trap (vacuum technology) A trap filled with a sorbant (adsorbant or absorbant material) to trap vapor.

Molecule A group of atoms held together by chemical bonds that has defined chemical properties. Often used in a context that includes atoms.

Molten salts (cleaning) Molten salts (chlorides, fluorides, borides) used for fluxing or metalliding. See Fluxing.

Molten salt electroplating Electroplating where the electrolyte is formed using molten salts (chlorides, fluorides) as the solvent. See Metalliding.

Momentum, particle A vector quantity equal to the mass (m) times the velocity (v) of the particle.

Monolayer (ML) A single layer of atoms or molecules on a surface in a close-packed arrangement.

Monomer A material consisting of simple molecular units (mers) that are capable of combining with other mers to form a polymer in which the monomer is a recognizable unit. See Polymer; Mer.

Morphology, bulk (film growth) The properties of the bulk of the film that can be visualized by fracturing the material, then observing the morphology of the fracture surface.

Morphology, surface (film growth) The properties of a surface such as roughness, porosity, long and short-range features, etc., that can be seen using an optical microscope, a scanning electron microscope (SEM), or an atomic force microscope (AFM).

Mother glass The glass substrate used in fabricating LCDs.

Mouse hole (film growth) Void left at the corner of the bottom of a trench during film deposition due to the top closing before the bottom is filled. Caused by geometrical shadowing. See Keyholing.

Movchan Demchishin (MD) diagram (film growth) Structure zone model of atomistically deposited vacuum condensates. See Structure zone models (SZM).

Mu metal (or Mumetal) A carefully annealed iron-nickel alloy (76 Ni, 16 Fe, 6 Cu, 2 Cr) that is used for magnetic shielding.

Multi-layer film (PVD technology) A film structure that contains two or more discrete layers of two or more different materials. Many layers can be formed by alternating deposition between vaporization sources. Examples: An X-ray diffraction grating of W-C-W-C-W; Ti-Pd-Cu-Au metallization. Also call a **Stack**.

Multi-stage vacuum pump (vacuum technology)   A vacuum pump with two or more stages in series within a single housing. See Vacuum pump.

Muratic acid   Another name for hydrochloric acid.

Mutagenic (chemical)   A chemical that has been shown to cause gene mutation in mice. See Carcinogenic.

N

NaK (contamination control)   An alloy of sodium (20-50%) and potassium that is liquid at room temperature and is used to getter oxygen and moisture in an inert gas dry box.

Nanoindentation (characterization)   Indentation of a surface using a very light load. Used to determine the hardness of a film.

Nanometer (nm)   A unit of length equal to 10^{-9} meters or 10 Ångstroms.

Nanoparticles   Clusters of several thousand atoms. Also called **Smoke, Soot, Nanoclusters,** or **Ultrafine particles.** See Gas Evaporation.

Nanophase material   Dense, ultrafine-grained material, often formed by atomistic vaporization processes, that has a high percentage (up to 50%) of its atoms at grain boundaries. Also called **Nanostructured material**.

Nanotechnology   The ability to create and utilize materials, devices, and systems through manipulation of matter on the nanometer scale.

Near-surface region (ion bombardment)   Region near the surface that is below the penetration region of the ions but which is affected by the bombardment by heating, diffusion, etc. See Altered region.

Near-surface region (surface analysis)   The region near the surface that is penetrated by the probing species or that generates the detected signal that is analyzed.

Nebulizer   Device for producing a fine spray of liquid droplets. Example: Ultrasonic nebulizer.

Negative glow region (plasma)   The bright region at the edge of the dark space in a DC glow discharge.

Negative ion   A particle that has one or more excess electrons.

Neutralization (electrical)   The removal of an electrical charge by the addition of charges of the opposite sign. Example: Surface charge neutralization. See Beam neutralization.

Neutralizer filament (ion gun, plasma source)   An electron emitting filament used to inject electrons into the ion beam that has been extracted from an ion gun, in order to eliminate "space charge blowup" of the ion beam.

Essentially changes the ion beam into a plasma beam.

Newton (N)   The SI unit of force.

Nichrome™ (material)   Trade name for the alloy 60Ni : 24Fe : 16Cr : 1C. Often used for metallization and for resistively heated wires.

Nickel sulfide (NiS) (inclusion)   An inclusion in glass that often causes fracturing when trying to fully temper the glass.

Nitric oxide (NO)   A good source of free oxygen that is easier to decompose than O_2.

Nitriding   Formation of a dispersion-hardened surface region by diffusion of nitrogen into a metal-alloy surface containing a material that will form a metal-nitride dispersed phase.

Noble metal   A metal whose atoms have satisfied electronic states and do not tend to react with other atoms. Example: Gold. See Noble gas; Noble Metal.

Noble species   An elemental species that has filled valence electron shells and thus is relatively chemically inert (e.g., He, Ne, Ar, Kr, Xe, Au). See Inert gas.

Nodule, film (film growth)   A visual mass of material that has a different appearance, microstructure, and/or morphology than the rest of the film material.

Non-aqueous cleaning   A cleaning procedure that does not need water during any portion of its use. See Semi-aqueous cleaning; Aqueous cleaning.

Non-aqueous electrolyte (electroplating)   An electrolyte formed by having a non-aqueous liquid solvent such as a fused salt or alcohol.

Non-aqueous electroplating   Electrodeposition of reactive materials such as aluminum using a non-aqueous electrolyte.

Non-destructive adhesion test (adhesion)   A test that can be performed to establish the presence of a specified amount of adhesion without destroying the film. Examples: Tape-test of a mirror surface; Pull-to-limit wire-bond test. See Adhesion test.

Non-linting material (cleaning)   A material that does not produce lint and is suitable for use in a cleanroom.

Non-permanent joints (vacuum technology)   Vacuum seals made so as to allow easy disassembly. The seal is made using an elastomer, a deformation metal seal, a shear gasket, or some other reusable or disposable material. See Permanent joints.

Non-polar molecule   A molecule that does not have any permanent electric dipole. Example: Oil.

Non-reactive deposition (film deposition)   Deposition where the material that is deposited is the same as the material that is vaporized. Usually performed in a vacuum or an inert gas environment.

Non-removable surfaces (vacuum technology)   The surfaces, such as chamber walls, that are not easily removed and must be cleaned in place. See Removable surface.

Normal glow discharge   A DC glow discharge in the pressure range that the current density on the cathode (**Cathode spot**) is constant with pressure changes. See Abnormal glow discharge.

Normal *(N)* solution (cleaning)   Solution containing one gram equivalent weight of material per liter of solvent. See Chemical solution, strength of.

Nozzle assembly (diffusion pump)   The arrangement of surfaces in a diffusion pump that gives the preferential direction to the vapors formed by heating the pump fluid. Also called **Jet assembly**.

Nucleation (film formation)   The stage of film formation where isolated nuclei are being formed on the substrate surface before the film becomes continuous.

Nucleation, de-wetting growth   When nuclei on a surface grow by adatoms avoiding the surface and the nuclei growing primarily normal to the surface. Example: Gold on carbon. See Nucleation; Wetting growth.

Nucleation, homogeneous   Uniform nucleation (nucleation density) over the whole surface.

Nucleation, inhomogeneous   Nucleation density that varies from place-to-place on the surface.

Nucleation, wetting growth   The lateral growth of nuclei on a surface due to the strong interaction of the adatoms with the surface. See De-wetting growth.

Nucleation density (film formation)   The number of nuclei per unit area on the substrate surface.

Nucleation sites, preferential (film formation)   Positions on a surface that have a high chemical reactivity and will react with mobile adatoms more readily than most of the surface. The site may be due to chemistry or morphology. Examples: Steps in the surface providing a high coordination at the base of the step; Inclusion of tin in one surface of float glass.

Nuclei, condensation (film formation)   The grouping of mobile atoms (adatoms) on a surface to form a stable structure. Stable nuclei can range in size from a few atoms (strong chemical bonding between the atom and the surface) to many atoms (weak interaction).

Nude gauge (vacuum technology)   A vacuum gauge that is inserted into the chamber volume and has no envelope or tubulation.

Number density (gas)   The number of gas molecules per unit volume.

"O" ring (vacuum technology)   An elastomer seal with a round cross-section. Used under compression as a vacuum seal. Also called an **O-ring.**

"O" ring, liquid (vacuum technology)   A ferrofluid seal used in rotary motion feedthroughs.

Oersted (Oe)   Unit of magnetic field intensity. Earth's magnetic field has a strength of about 0.5 Oe. See Gauss.

Off-cut surface (substrate)   See Vicinal surface.

Off-plating (electroplating, cleaning)   The removal of material from the anode in an electrolysis cell.

Ohm (characterization)   A unit of electrical resistance. See Sheet resistivity.

Ohm-centimeter (ς-cm)   A unit of bulk electrical resistivity (r). Example: The resistance R, in ohms, of a wire having a length L, a resistivity of r, and a cross-sectional area of A is given as $R = rL/A$.

Ohmic contact (metallization)   A low-resistance, non-rectifying electrical contact between a film and a substrate.

Ohms-per-square (characterization)   Resistivity unit used for thin film structures. See Sheet resistivity.

Oil mist accumulators (vacuum technology)   A trap to prevent the loss of oil through the exhaust system. Also called an **Exhaust trap**; **Demister**.

Oil-free vacuum pump (vacuum technology)   Vacuum that doesn't use oil for sealing or lubrication in a way that might contaminate the processing chamber. Also called a **Dry pump**. See Vacuum pump.

Oil-sealed vacuum pump (vacuum technology)   A vacuum pump that uses oil to seal the space between moving surfaces.

Oleophilic wick (cleaning)   An oil-loving fabric used to skim oil from surfaces.

Open, electrical (semiconductor technology)   Where a portion of an electrical conductor stripe is missing. Detectable by voltage-contrast techniques in an SEM.

Open porosity (substrate)   Interconnected pores that provide a path from the interior of the material to the surface. See Closed porosity.

Operational spares (vacuum technology)   Spare parts to replace parts which, if they fail or need to be replaced for any reason, will prevent use of the equipment. Examples: Spare roughing pump; Spare O-rings.

Operator (manufacturing)   The person operating the equipment, performing the process, or implementing the MPIs. See Training, on-floor; Training, formal.

Ophthalmic coatings   Coatings on eyewear such as sunglasses.

Optical adsorption spectroscopy (process control)   Characterization of a gaseous medium by measuring the adsorption of a spectrum of radiation as it passes through the gas or vapor. Characteristic wavelengths are adsorbed by the gas and the amount of adsorption depends on the number density of atoms along the pathlength. Can be used as a vaporization rate monitor.

Optical coating(s) (optics)   Single and multilayer film structures used to obtain desired transmittance and reflectance of radiation from surfaces. The property may be due to the intrinsic property of the material (e.g., an aluminum reflector) or due to interference effects. A multilayer optical coating is also called an **Optical stack**.

Optical coating(s) (decorative, security)   Single and multilayer film structures used to obtain desired visual effects such as color, texture, light scattering, etc.

Optical coating(s), active   Film structures that change optical properties under an external stimulus.

Optical density (OD) (characterization)   The logarithm of the ratio of the percent of visual light transmitted through the substrate without metallization, to the percent of visual light transmitted through the metallized substrate. Example: 1% transmission is an OD = 2.

Optical emission (plasma)   The emission of radiation from a plasma due to de-excitation of excited species.

Optical emission spectroscopy   Technique of measuring the optical emission from a plasma. Used to determine the species and density of particles in a plasma.

Optical pyrometry   Determination of the temperature of a surface by observing its color temperature, usually by comparing its color to the color of a surface at a known temperature. See Infrared pryometry.

Optical spectrum   The visible and near-visible wavelengths. The extreme limits are taken as 0.1 micron in the ultraviolet and 30 microns in the infrared. See Visible radiation.

Optical thickness (optics)   The product of the physical thickness and the index of refraction of the material.

Optically stimulated electron emission (OSEE) (cleaning)   Electron emission from a metal surface under ultraviolet light radiation. Changes in OSEE can be used to quantify surface contamination.

Optically variable device (OVD)   A device that presents a different picture when viewed from different angles. Often used as a security measure.

Orange peel (surface)   A uniformly rough, pebbly-looking surface morphology that resembles the surface of an orange. Often seen on smooth polished surfaces or cured polymer surfaces.

Orbital welder (semiconductor equipment)   An automated arc welder that is used to weld the stainless-steel tubing in gas distribution systems. See Material Test Report (MTR).

Organic material   Material consisting of mostly hydrogen and carbon.

Orifice, ballast (vacuum technology)   An opening that continuously allows gas from the outside to bleed into the foreline of a pumping system. This prevents suckback in the case of a power failure. By using dry air into the orifice, moist air is diluted to the point that water vapor is not condensed by compression in the mechanical pump.

Original equipment manufacturer (OEM) (manufacturing)   The outside supplier of processing equipment that conforms to certain specifications. The supplied equipment may be modified to meet special requirements in the manufacturing environment. See Beta test.

Ounce (metal sheet)   Measure of sheet thickness by giving the weight of the sheet per square foot. Example: 16 ounce copper sheet. See Gauge.

Ounce, fluid   A measure of fluid volume equal to $1/32$ of a quart. Often just called an ounce.

Outdiffusion (cleaning)   The diffusion of a species from the bulk of a material. Often used to describe mobile materials that do not vaporize when they reach the surface.

Outgassing (vacuum technology, cleaning)   The diffusion and volatilization of species from the bulk of a material.

Outgassing rate   The amount of gas leaving a surface as measured by Torr-liters/sec-cm^2.

Over-diffusion (adhesion)   When the extent of the interdiffusion of materials causes a weakening of the material in the diffusion zone. Examples: Weakening by formation of Kirkendall porosity; Microfracturing due to stresses caused by phase changes in the diffusion zone.

Over-flow rinse tank (cleaning)   Tank containing rinse water that flows off the top to carry away contaminants that float on the surface. This prevents "painting-on" of the contaminants onto the surface as the surface is withdrawn from the tank. See Counter-flow rinsing.

Overlay coatings   Coatings formed by the addition of another material to the substrate surface. The original substrate material is not detectable in the coating. See Surface modification.

Oxidation, chemical (cleaning)   Loss of electrons, typically by reaction with oxygen, chlorine, fluorine, or bromine.

Oxidation cleaning (cleaning)   Removal of contaminant species by oxidation and solution or volatilization.

Oxidizing agent (cleaning)   A material that causes oxidation and is thereby reduced.

Oxygen plasma cleaning (cleaning)   Cleaning in an oxygen plasma where the contaminant is oxidized and vaporized.

Oxygen transmission rate (OTR)   The amount of oxygen transmitted through a film in units of amount per unit area per unit time and pressure differential. Units of STD cm^3/m^2-d-bar

Ozonated (cleaning)   Ozone added. Example: Ozonated de-ionized water.

Ozone (cleaning)   The molecular form of oxygen, O_3, which is very chemically reactive. Generated in large amounts in a corona or arc discharge at atmospheric pressure. Generated in smaller amounts in short-wavelength ultraviolet radiation and in low-pressure oxygen glow discharges. Used for cleaning.

Ozone cleaner (cleaning)   Gaseous cleaning technique that uses ozone to produce volatile oxidation reaction products such as CO and CO_2 from the oxidation of hydrocarbon contaminants. Also called **UV/O_3 cleaner**.

Ozone depletion potential (ODP) (cleaning)   A rating for the potential of a vapor to deplete the atmospheric ozone layer. See Global warming potential (GWP).

P

Pack cementation (CVD)   A CVD-type process where the part to be coated is placed in a mixture (**Pack**) of inert powder and powder of the material to be deposited. The mixture is heated and a reactive gas reacts with the coating powder to form a chemical vapor precursor that decomposes and diffuses into the surface of the part. Used to carburize, aluminize, and chromize surfaces.

Packaging   A protective or containing unit used to enclose a material. Examples: A metallized plastic bag for potato chips to contain the chips and prevent moisture from combining with the salt on the chips; Ceramic container used to isolate a semiconductor device from the ambient.

Paddle (semiconductor processing)   The tooling that slides under and picks up the silicon wafer.

Paramagnetic   A material in which an applied magnetic field will produce magnetization in the same direction (positive magnetic susceptibility), but has no magnetic moment of its own. Most non-magnetic materials are paramagnetic.

Parameter window (manufacturing)   The limits to a process variable, such as temperature, between which an acceptable product will be produced.

Partial pressure (vacuum technology)   The pressure of a specific gas or vapor in a system. See Dalton's Law of Partial Pressures.

Partial pressure analyzer (vacuum technology, reactive deposition)   A device, such as a mass spectrometer or an optical emission spectrometer, that is used to determine the partial pressure of each gaseous species in a gas mixture.

Particle, fine (cleaning)   A particle whose diameter is less than 2.5 microns (EPA definition).

Particle, respirable (safety)   A particle small enough to be inhaled into the lung of a healthy person, usually smaller than 10 microns.

Particle, ultrafine (cleaning)   Particle having a diameter less than about 0.5 micron. Generally formed by vapor phase nucleation or the residue from the evaporation of an aerosol. See Vapor phase nucleation.

Particulate contamination (cleaning)   Contamination by particulates. A major source of pinholes in thin films either by geometrical shadowing or by holes generated when the particle is dislodged from the surface.

Parting layer   See Release layer.

Parylene process   A polymer film deposition process where a monomer is passed through a heated zone where it is polymerized and the resulting polymer (Example: polyparaxylyene) is then condensed onto a surface under very benign conditions.

Pascal (Pa)   A unit of force equal to a Newton per square meter. 6,900 Pa (6.9 kPa) = 1 psi. See Pressure, units of.

Paschen curve   The curve of the breakdown voltage as a function of the product of pressure (p) times the separation (d) (i.e., p x d) for two electrodes (Rojowski-shaped) in a low-pressure gas.

Pass box (contamination control)   Two-door container mounted in a wall that allows passing items from one room to another in a controlled manner.

Passivation   Producing a surface layer on a material that decreases its reaction with the ambient. Passivation can be accomplished by removing a reactive species from the surface (ASTM A380), increasing the thickness or density of a naturally forming oxide, chemically reacting the surface to form a passive compound, or by overlay techniques where the passivation layer consists of a different material.

Passive film   A film that does not change properties under stimulation. Example: Aluminum mirror coating. See Active film.

Passive storage (cleaning)   Storage in an environment that has been cleaned in the past but is not actively being cleaned during the storage. See Active storage.

Patent, provisional   A temporary patent that establishes a file date for the disclosure. The provisional patent expires at the end of one year, at which point a utility patent with disclosures and claims should be filed.

Patent, utility   A document issued by the U.S. Patent and Trademark Office (USPTO) that grants exclusive use of a process, product, or composition of material in the United States to the holder of the patent for a period of 20 years after the first filing date.

Patina   Term used to describe the weathered look of a metal such as the dark green patina formed on weathered copper. The color of the patina often depends on the composition of the weathering environment.

Penning ionization (plasma)   Ionization of an atom by collision with a metastable atom in an excited state that is of higher energy than the ionization energy of the first atom. Example: Ionization of copper (ionization energy = 7.86 eV) by excited argon (metastable excited states of 11.55 and 11.75 eV).

Penning vacuum gauge (vacuum technology)   An ionization vacuum gauge in which the electric and magnetic fields are approximately parallel. Also called the **Phillips ionization gauge**. See Vacuum gauge.

Percent solution (solution strength)   The percent, by weight, of a pure chemical in water. See Chemical solution, strength of.

Perfect gas   A gas that is composed of atoms or molecules that physically collide but otherwise do not interact. Low-pressure gases are generally treated as ideal gases. **See Ideal gas** (preferred).

Periodic deposition (film formation)   1) When a film of a compound material is formed by periodically depositing a very thin metal film, then reacting the film with a gaseous reactant, then repeating the process many times to build up the film thickness. Example: The MetaMode™ deposition process. 2) When a multilayer film structure is formed by depositing many alternating layers of different materials using a fixture that exposes the substrate to first one, then the other material. Example: Alternating layers of tungsten and carbon for an X-ray diffraction grating. Also called **Alternating plating.**

Peristaltic pump (CVD)   A liquid pump that operates by creating a wave motion, by constriction and expansion, in a tube carrying the fluid or by a moving diaphragm.

Permanent joints (vacuum technology)   A vacuum seal that is made so as to not be disassembled easily. Examples: Weld joint; Braze joint. See Non-permanent joints.

Permeation   The passage of a gas or vapor through a solid barrier. See Diffusion.

Permeation rate   Permeation measured in Torr-liters/sec-cm^2 or grams/sec-cm^2.

Permissible exposure limits (PEL) (safety)   Permissible exposure limits to hazardous materials (OSHA). See Time-weighted average (TWA); Short term exposure limits (STEL).

Pennyweight   Unit of weight in the Troy Weight System equal to 24 grains or 1.555 grams.

Perchloroethylene (PERC) (cleaning)   The solvent perchloroethylene (CCl_2CCl_2).

Pewter (metal alloy)   An easily castable, tin-based, non-workhardenable alloy. A non-leaded composition is 91% tin, 7.5% antimony, and 1.5% copper. A leaded composition can be 70% tin and 30% lead.

pH (Pouvoir hydrogene)   The logarithm of the reciprocal of the H^+ ion concentration of a solution. Very pure water at 22°C has an H^+ ion content of 10^{-7} moles per liter, i.e., a pH of 7. A concentration of 0 to 7 is acidic (e.g., a 1 molar HCl solution has a pH of 0; a 0.1 normal H_2SO_4 solution has a pH of 1.17) and 7 to 14 is alkaline or basic (e.g., a 1 molar NaOH solution has a pH of 14, a 0.1 normal NH_4OH solution has a pH of 11).

Phase   A physically identifiable region of material that has specific characteristics. There are solid, liquid, and gaseous phases as well as various solid phases that can be characterized by crystalline structure or by chemical composition.

Phase, crystalline (crystallography)   A physically distinct state of matter or portion of matter (grain, crystallite, inclusion, etc.) that can be defined by analytical means (X-ray diffraction, transmission electron microscopy, etc.).

Phase, thermodynamic   The state of matter such as a solid, a liquid, or a gas.

Phase change   The changing from one phase to another due to compositional, temperature, or pressure changes.

Phase diagram   A diagram showing the phases of a material or a mixture of materials as a function of temperature and/or pressure and/or composition.

Phosgene   A toxic gas with the formula $COCl_2$.

Phosphate conversion (surface modification)   The production of an electrically-conductive metal phosphate on the surface of a metal by wet chemical reaction. Example: Use of zinc or manganese acid phosphate treatment of aluminum for corrosion protection. See Chromate conversion.

Phosphor   A material that converts an impinging particle radiation, such as electron bombardment, into optical radiation. Example: Cathode ray tube (CRT).

Photodensitometer (characterization)   An instrument for determining the areal densities of an image on a photograph. For example: Used to determine particle size distribution in an abrasive powder. See Densitometer.

Photodesorption   The desorption of species from a surface due to heating by resonant adsorption of the incoming radiation.

Photoelectron emission   Electron emission stimulated by the resonant adsorption of electromagnetic radiation. Example: Photoelectric effect.

Photoexcitation   Excitation of an atom or a molecule by resonant adsorption of incident radiation.

Photoionization   Ionization of an atom or molecule by resonant adsorption of incident radiation.

Physical sputtering (PVD technology)   Often called just **Sputtering**. The physical ejection (vaporization) of a surface atom by momentum transfer in the near-surface region by means of a collision cascade resulting from bombardment by an energetic atomic-sized particle.

Physical vapor deposition (PVD)   The deposition of atoms or molecules that are vaporized from a solid or liquid surface. See Chemical vapor deposition (CVD).

Physisorption   The retaining of a species on a surface by the formation of weak chemical bonds (< 0.2 eV) between the adsorbate and the adsorbing material. Also called **Physical adsorption**. See Chemisorption.

Pickling (cleaning)   Removal of large amounts of a surface layer, such as an oxide scale, by chemical means. Example: Acid pickling.

Pick-n-place (semiconductor processing)   A robotic motion to take a wafer from one position and place it in another. Example: From cassette-to-cassette.

Pigment   Material added to a paint or ink to produce a color or an optical effect. Example: Particles derived from an optical interference stack to produce angle-of-incidence color changes in a paint.

Pilot production   Production to evaluate a process flow using full-scale equipment or equipment that can be scaled-up to meet production throughput requirements.

Pinhole (film formation)   A small hole in the film due to incomplete coverage during film growth or from flaking (**Pinhole flaking**). See Porosity, film.

Pinhole flaking (contamination control)   Flaking from film build-up on surface aspirates producing particulate contamination in the deposition system.

Pipe diffusion (semiconductor technology)   Rapid diffusion along a dislocation.

Piranha solution (cleaning)   An oxidative cleaning solution based on sulfuric acid and ammonium persulfate. Used to clean silicon wafers.

Pirani gauge (vacuum technology)   A vacuum gauge that uses the resistance of a heated resistor element, which can change due to gas cooling, as an indicator of the gas pressure (density). See Vacuum gauge.

Piston pump   A positive displacement vacuum pump that uses the motion of a piston(s) to compress the gas.

Planar magnetron (sputtering)   A magnetron configuration where the target surface is a planar surface and the magnetic field is in a configuration such that the oval sputter-erosion track resembles a "racetrack." See Magnetron.

Planarization (semiconductor processing)   To smooth a surface, generally by polishing, after filling a via with metallization.

Plasma   A gas that contains an appreciable number of electrons and ions such that it is electrically conductive.

Plasma generation region   The region in which free electron and ions are generated.

Plasma, augmented   A plasma whose electron density has been increased by the addition of electrons from an external electron source such as a hollow cathode.

Plasma, auxiliary   A plasma separate from the main processing plasma. For example, an auxiliary plasma is needed near the substrate to activate the reactive gas in reactive magnetron sputtering where the main plasma is confined away from the substrate.

Plasma, equilibrium   A plasma that is volumetrically neutral.

Plasma, low-density   A plasma that has a low particle density.

Plasma, strongly ionized   A plasma where most of the gaseous particles are ionized.

Plasma, weakly ionized   A plasma in which only a small percentage (e.g., 0.01%) of the gaseous particles are ionized and the rest of the particles are neutral.

Plasma activation (film formation)   Making gaseous species more chemically reactive in a plasma by excitation, ionization, fragmentation or by the production of new chemical species. See Reactive deposition.

Plasma anodization   Oxidation of an anodic surface in contact with a plasma containing oxygen.

Plasma assisted CVD (PACVD)   See Plasma Enhanced CVD (PECVD).

Plasma bucket   A multipole magnetic field arrangement used to spread the plasma emerging from a plasma source into a downstream processing region.

Plasma cleaning (cleaning)   Cleaning using a plasma environment. The cleaning action can be from desorption (inert gas plasma) or chemical reaction and volatilization (reactive gas plasma).

Plasma-compatible materials (plasma technology)   Materials that do not change properties in the presence of a plasma and do not contaminate the plasma. Many organic polymers are not plasma compatible due to their degradation by the UV from the plasma.

Plasma deposition   Formation of a film by the use of a plasma to decompose or polymerize a precursor gas or vapor. See also Plasma polymerization; Plasma Enhanced Chemical Vapor Deposition (PECVD).

Plasma duct (arc vaporization)   A filtered arc source where the plasma is magnetically deflected so that the macros are deposited on the wall of the duct. See Arc source.

Plasma enhanced CVD (PECVD)   Chemical vapor

deposition where a plasma is used to assist in the decomposition and reaction of the chemical vapor precursor, allowing the deposition to be performed at a significantly lower temperature than when using thermal processes alone. Example: PECVD of phosphosilicate glass (PSG) encapsulating glass at 450°C in semiconductor processing. See Reinberg reactor; Low pressure CVD (LP-CVD).

Plasma etcher (semiconductor processing) A vapor etching system that uses a plasma to activate the etchant vapor, which then reacts with a surface to form volatile reaction products. Examples: BCl_3 plasma etching of aluminum; CF_4 plus O_2 plasma etching of silicon.

Plasma immersion ion implantation (PIII) A process in which a metallic substrate is immersed in a plasma and pulsed momentarily to a high potential (50-100 kV). Ions are accelerated to the surface from the plasma and before there is an arc-breakdown, the pulse is terminated.

Plasma parameters (plasma technology) Important plasma parameters are: electron density, ion density, ion charge state distribution, density of neutral species, electron temperature, ion temperature, and average particle temperature. Uniformity of the plasma parameters from place-to-place in the plasma can be important in plasma processing.

Plasma polymerization The conversion of a monomer vapor to a polymeric species in a plasma or on a surface exposed to a plasma. The monomer may or may not be recognizable in the resulting polymer.

Plasma potential The potential of the plasma with respect to a surface in contact with the plasma that may be grounded, floating, or electrically insulating. The plasma potential will always be positive with respect to any large-area surface that it is in contact with.

Plasma source (plasma technology) A device for generating a plasma. Often a plasma beam is formed using an electron emitting source in a magnetic and electric field. In some cases a plasma beam is formed from an ion beam by adding enough electrons to produce volume neutralization.

Plasma source, capacitively coupled rf A plasma source where the plasma is formed in a region between two parallel-plate electrodes driven by rf power. See Reinberg reactor.

Plasma source, electron cyclotron resonance (ECR) A plasma source where the microwave energy, which has a resonant frequency of the electron in a magnetic field, is injected into the plasma-generating region through a dielectric window.

Plasma source, gridless end-Hall A plasma source that uses a thermoelectron emitter and a magnetic field to confine the electrons so as to impinge on gas molecules exiting an orifice.

Plasma source, helicon A plasma source in which microwave power is used to accelerate electrons in a gas in the presence of a constant magnetic field.

Plasma source, inductively coupled plasma (ICP) A plasma source where the plasma is formed in a region surrounded by an rf coil that couples energy into the electrons in the plasma.

Plasma spraying Melting small particles in a high-enthalpy plasma and a high-velocity gas stream (1200 ft/sec) and "splat cooling" them on a surface. Plasma spraying is a type of **Thermal spray processing**.

Plasma-based ion plating Ion plating where the substrate is in contact with a plasma. Typically ions are extracted from the plasma to bombard the substrate and growing film. The plasma also activates reactive gases in the plasma during reactive ion plating. See Ion plating.

Plasma-deposited films Films deposited from a plasma using a chemical vapor precursor gas or a monomer as a source of the deposited material. See Plasma polymerization; Plasma enhanced CVD; Chemical ion plating.

Plastic deformation The permanent deformation of a material under a mechanical stress that exceeds its elastic limit.

Plasticizer (contamination) A low-molecular-weight, generally organic material added to polymer resins to make them more fluid and moldable. Plasticizers can be a major source of contamination coming from the bulk of a molded polymer material.

Plate glass Flat, high-quality glass with plane parallel sides. Usually formed by the float glass process. May be bent into a curvature as, for example, automotive windshields. May be tempered or hardened. Also called **Flat glass.**

Platinum black Ultrafine particles of platinum often made by gas evaporation. Used as a catalyst.

Plug (metallization, semiconductor processing) The material filling a hole or via in the structure. Example: CVD tungsten plug.

Plume (laser) The cloud of vapor that rises from the heated spot during laser vaporization. The cloud adsorbs some of the laser radiation to produce ions and electrons

Pocket (e-beam evaporation) The cavity in the water-cooled copper hearth that holds the material to be evaporated in electron beam evaporation. See Liner.

Point-of-use (manufacturing) The point in the processing flow that the material will be used. Example: Measuring the electrical conductivity of ultrapure water distributed though a manifold system at the point it will be used.

Poisoning, target (sputtering) Reaction of the surface of a sputtering target either with the reactive gas being used for reactive deposition or with a contaminant gas. The reacted layer causes a change in the performance of the sputtering target.

Poisson's ratio The ratio of the contracting strain in the diameter direction to the elongation strain in the axial direction when a rod is pulled in tension.

Polar molecule (cleaning) A molecule that has a permanent electric dipole. Example: Ionic salts. See Nonpolar molecules.

Polarity (electricity) An indication of which direction an electron will flow. An electron will flow away from a negative pole (cathode) toward a positive pole (anode) if there is a potential difference between the two electrodes.

Polarization The process of producing relative displacement between positive and negative charges.

Polarization bonding Chemical bonding due to polarization of two atoms or molecules. Also called **van der Waals bonding.** See Chemical bond.

Polished brass coating (decorative coating) The zirconium nitride coating deposited by reactive PVD processes that resembles polished brass in color.

Polishing, chemical (surface modification) Increasing the surface smoothness by using a chemical etch that preferentially removes high spots on the surface. Examples: Polishing aluminum in 10% HCl; Polishing stainless steel in a mixture of acids.

Polishing, electropolish (surface modification) Polishing a surface that is the anode of an electrolysis cell using a suitable electrolyte. Example: Electropolishing stainless steel in a phosphoric acid-based electrolyte.

Polishing, mechanical (surface modification) The use of abrasives of varying sizes to mechanically abrade a surface to increase surface smoothness.

Polishing, of water (cleaning) Taking ultrapure water that has been used in processing and sending it back through the water purification system by injecting it downstream of the initial stages of purification.

Polishing compound A material used to smooth a surface or to give the surface a specific texture. Removal of surface material is a secondary consideration. Examples: Cerium oxide; Chromium oxide; Diamond. See Abrasive compound.

Poly (semiconductor processing) Slang term for a polysilicon film.

Polyamide (substrate) A condensation-type polymer. Polyamides can retain large amounts of water. Example: Nylon™.

Polycold (vacuum technology) A term sometimes used for a cryocondensation (Meissner) coil/panel in the process chamber. Polycold is the company that popularized such traps in the USA.

Polyethylene terepthalate (PET) (substrate) A poly-

mer material used for webs and plastic containers. PET film is a biaxially oriented material that has good transparency, toughness, and permeation barrier properties. Example: DuPont Mylar™.

Polyimide (substrate) A high-temperature polymer. Example: Kapton™.

Polymer A material comprised of giant molecules formed by the chemical bonding of small chemical units called **mers.** The bonding may form a **Linear chain** or there may be multiple bonds between monomers to form highly "**Crosslinked**" polymers. See Copolymer.

Polypropylene (PP) (substrate) A polymer material that is used for webs and plastic containers. Less expensive than PET but has less desirable optical properties.

Polysilicates Three-dimensional polymer of Si-O; i.e., essentially every silicon atom is bonded to four oxygen atoms.

Polysilicon (semiconductor technology) A film of polycrystalline silicon. Also called **Poly.**

Polysiloxanes Three-dimensional polymer of Si-O except that 5-10% of the silicon atoms are bonded to one hydrocarbon moiety, usually a methyl or phenyl group.

Polysilsesquioxanes Three-dimensional polymer with the formula $(RSiO_{1.5})_n$; i.e., every silicon atom is bonded to one hydrocarbon moiety, usually a methyl group or a combination of methyl and phenyl groups.

Porosimetry Determination of the open pore volume in a material. Example: Mercury porosimetry where mercury is hydrostatically forced into the pores and the weight-change measured. Porosimetry can be used in the specification of sputtering targets formed by powder pressing processes.

Porosity, open Pore volume that is interconnected and connected to the surface. May or may not affect measured density depending on the measuring technique.

Porosity, film Open or closed porosity in the deposited film due to the mode of growth, substrate effects, void coalescence, or pinhole flaking. See Columnar morphology; Macrocolumnar morphology.

Porosity, closed Pores that are not connected to the surface. Affects density measurements.

Porous silicon A network of nano-sized silicon regions surrounded by void space. Prepared by electrochemical anodization of a silicon surface.

Port, vacuum An opening through a chamber wall into the vacuum chamber. See Flange.

Position equivalency When all positions on a fixture yield parts that are indistinguishable one from another or that lie within an acceptable range of property variation. If position equivalency is not established, the batch can have unacceptable variations in the properties of the coated parts.

Positive column (plasma)   The field-free, luminous region in a DC gas discharge between the negative glow and the anode. The region that allows the use of gas discharges for linear illumination.

Positive displacement vacuum pump   A mechanical vacuum pump that traps a volume of gas, compresses it, and displaces it through an exhaust port. See Vacuum pump.

Post magnetron (sputtering)   A magnetron configuration that is a post, perhaps with flares on the ends (**Spool**), with a magnetic field either axial to the post or in a series of looped magnetic fields around the post. The electrons are confined along the surface of the post and between the flared ends. See Magnetron.

Postdeposition treatments (film formation)   Treatments to change the properties of the film after deposition. Examples: Topcoating; Shot peening; Burnishing to close porosity.

Postvaporization ionization (PVD technology)   Ionization of the vaporized (sputtered or evaporated) film atoms to form **Film ions** that can be accelerated in an electric field. See Film ions.

Potential (electrical)   The voltage at a position generally with respect to ground. See Voltage.

Potential (energy)   The energy of a body due to its position in a force field such as gravity. For example, the kinetic energy a person will acquire if they fall off a ladder. See Kinetic energy; Momentum.

Powder coating (substrate)   Coating formed by the deposition of a powder by spraying or electrostatic spraying, generally followed by heating to fuse the particles together and to the surface. The Powder Coating Institute's <u>Powder Coating Manual</u> describes the techniques used.

Power, target (sputtering)   The power (watts) or power density (watts/ cm^2) applied to the sputtering target. This process variable, along with gas pressure and gas composition, are the parameters most often used to control the sputtering and sputter deposition processes.

Precision   The closeness of agreement between randomly selected individual measurement or test results. See Repeatability; Accuracy.

Precision cleaning   I know of no better definition than "Cleaning what already looks clean." Also called **Critical cleaning.**

Precursor, chemical, liquid (CVD, PVD reactive deposition)   A liquid that acts as the source of the depositing material by containing the elemental constituents of the coating that are released by heating, reduction, etc. The liquid is vaporized in a hot chamber and carried into the deposition chamber by a hot **Carrier gas.** Example: TiCl$_4$ whose boiling point (bp) is 136.4°C as a source of titanium.

Precursor, chemical, vapor (CVD, PVD reactive deposition)   A vapor (at room temperature) that acts as the source of the depositing material by containing the elemental constituents of the coating that are released by heating, reduction, etc. Examples: SiH$_4$ as a source of silicon; C$_2$H$_2$ as a source for carbon.

Preferential evaporation   When one constituent of an alloy vaporizes faster than another because of its higher vapor pressure at a specific temperature.

Preferential nucleation sites (film growth)   Positions on a surface where the mobile adatoms prefer to condense. Examples: Charge sites; Atomic steps; Interfaces; Lattice defects such as grain boundaries; Substitutional atoms; Emerging dislocations.

Preferential sputtering (surface composition)   When one constituent of the surface sputters more rapidly than another, leaving a detectable **Surface enrichment** of the low-sputtering-yield material. Note that this layer must be sputtered before the underlying material is exposed so the ratio of the constituents in the deposit is the same as that of the bulk material, even though there is surface enrichment.

Preferential sputtering (surface morphology)   When one area sputters faster than another because of an inclusion, grain orientation, or other factors. Leads to roughening of the surface or a delineation of certain types of surface features. See Cones.

Preferred orientation (crystallography)   When non-random growth gives the film microstructure a preferred crystal orientation (texture) in some plane.

Premelting (evaporation)   Melting the evaporant charge while the shutter is closed. This allows degassing of the charge and establishes good thermal contact of the heated surface to the charge material before the shutter is opened and deposition begun.

Presputtering, target (sputtering)   Sputtering a target with a shutter closed or with the substrates out of line-of-sight, to clean the surface of the target. Also called **Target conditioning.**

Pressure, base (vacuum technology)   A specified pressure for the system to begin the first sequence in the processing. See Pumpdown time.

Pressure, blank off (vacuum technology)   The lowest pressure that a vacuum pump can attain at the inlet side when the inlet has been blanked off.

Pressure, gas (vacuum technology)   The force per unit area exerted by gas molecules impinging on a surface. See Pressure units.

Pressure, ultimate (vacuum technology)   The pressure in a system toward which the pumping curve seems to be approaching asymptotically, under normal pumping and processing conditions. Value will never be reached and depends on the sources of gases and vapors in the system. See **Base pressure.**

Pressure, units of (vacuum technology)   The units of force per unit area used to measure gas pressure. It is important in communication to make sure that each individual knows in what pressure units the other person is talking. Example: "We established the plasma at 10^{-3}" (Torr, mbar, Pascals?).

Pressure, units of, bar   One bar of pressure equals 10^5 Pascals. 1 bar = 0.98692 atmospheres = 750.06 Torr. The bar and millibar are pressure units commonly used in Europe. A **millibar (mbar)** is one-thousandth of a bar.

Pressure, units of, Pascals   A unit of pressure equal to a Newton per square meter. 6900 Pa (6.9 kPA) = 1 psi.

Pressure, units of, pounds-per-square-inch (psi)   A unit of pressure equal to one pound per square inch.

Pressure, units of, Torr (or torr)   A unit of pressure defined as $1/760$ of a standard atmosphere. A **milliTorr (mTorr)** is one-thousandth of a Torr.

Pressure, working (processing)   The pressure at which the process is being performed.

Preventive maintenance (PM)   Periodic maintenance performed to reduce unexpected failure of equipment and extend the life of the equipment. This is opposite of **"Run-to-crash"** approach. Example: Periodic oil (lubricant or sealant) change.

Primary standard   A unit whose value (e.g., leak rate, resistivity, length, composition) has been established by an accepted **Authority** (e.g., NIST in the USA) against which other units are calibrated. Generally the primary standard must be periodically recalibrated by the authority. See Secondary standard.

Printed circuit (PC)   A conductive pattern on an insulating surface that may or may not include active devices such as relays (large) or semiconductor devices (small). If semiconductor devices are applied to the circuit pattern (**Appliquéd**), the circuit is called a **Hybrid microcircuit.**

Privacy filters (optics)   Optical structures that provide severe angle-dependent transmission characteristics such that observers to the side cannot read a screen.

Process flow diagram (PFD)   A diagram showing each successive stage in the processing including storage, handling, and inspection. A PFD is useful in determining that there are MPIs that cover all stages of the processing.

Process parameter window   The limits for each process parameter between which a good product is produced. See Robust process.

Process parameters   The variables associated with the process that must be controlled in order to obtain a reproducible process and product. Examples: Time; Temperature; Target power; Gas pressure; etc.

Process review meetings (manufacturing)   Periodic meetings of engineers from the various shifts, managers, and persons involved in developing the specifications, to review changes to the specifications and MPIs and to discuss other matters affecting product yield, throughput, quality, etc.

Process sheet   The process sheet that details the process parameters of the deposition run. Also called a **Run sheet.** See Traveler.

Product throughput   The number of units produced per unit time.

Profilometer, surface   Instruments for measuring the surface morphology and roughness.

Properties, film   Properties of the film that are determined by some specified technique.

Properties, film, functional   Properties that are essential to the desired function of the film such as sheet resistance for conductivity, optical reflectance for mirrors, etc.

Properties, film, stability   Properties that influence long-term performance such as corrosion resistance, residual film stress, etc.

Pseudodiffusion-type interface (film formation)   An interfacial region where the material is graded, similar to the diffusion interface, produced by mechanical means such as beginning the second deposition before stopping the first deposition, or by implantation of high-energy "film ions."

Pseudomorphic structure   A crystalline structure that has been altered by stress, solute atoms, etc.

Pull-outs (adhesion)   Regions of the film having poor adhesion, which are pulled out by adhesion tests (tape test, stud-pull test, etc.), leaving pinholes.

Pull-outs (surface)   Areas on a surface that are easily removed, leaving craters in the surface.

Pulse plating (electroplating)   The use of a pulsed DC for plating rather than a continuous DC. This allows higher momentary current densities that can affect the coating morphology. In some cases the polarity may be reversed to give **Off-plating** of the part, which affects the coating morphology. See Off-plating.

Pulse power   Processing power that is only applied for a portion of the total process cycle time. Examples: Pulsed DC; Pulsed bipolar.

Pulsed DC   A DC waveform that has a voltage that is less than the cycle time and the rest of the cycle being at zero potential. See Direct current.

Pulsed laser deposition (PLD)   Deposition using laser ablation as the vaporization source. See Laser vaporization.

Pump, direct-drive (vacuum technology)   A mechanical pump where the moving parts of the pump are connected to the motor by a rigid shaft (no belt).

Pump capacity (vacuum technology)   The amount of a specific gas that a capture pump, such as a cryopump, can contain and still pump effectively. When this value is exceeded, the pump must be regenerated. See Regeneration.

Pumpdown time   The time for a vacuum system to reach a specified pressure (base pressure).

Pump package (vacuum technology)   A combination of pumps designed to work together in sequence. Examples: Diffusion pump - Roots blower - oil-sealed mechanical vane pump; Turbopump - molecular-drag pump - diaphragm pump. Also called **Pumping stack.**

Pump throughput (vacuum technology)   The mass of gas (or number of molecules of gas) that pass through a pump per unit time (Torr-liters/sec). Also called **Mass throughput.**

Pumping, full flow (vacuum technology)   Pumping on the processing system with as high a conductance as possible.

Pumping, throttled (vacuum technology)   Pumping on a processing system with some conductance restriction in place to lower the actual pumping speed of the system.

Pumping speed   The volume flow rate through a vacuum pump in liters per second. Also called **Pump speed**. See Mass throughput; Pump throughput.

Pure water (cleaning)   Water formed by reverse osmosis filtration of ions and particulates, activated carbon filtering of organics, and mechanical filtering of particulates and living organisms (wee beasties). Often used as a final rinse when ultrapure water is not required. See Ultrapure water.

Purge (vacuum technology, semiconductor processing) To flow a gas (**Purge gas**) through a system to displace and remove gases, vapors, and loose particulates that are present.

Purple plague (adhesion)   The color of the fracture surface in an Au-Al interface when the intermetallic Au_2Al is formed.

Pyrolysis   The fragmentation of heavy molecules by heat.

Pyrophoric gas   A gas that will spontaneously ignite if exposed to air at or below 54°C (130°F). See Flammable gas.

Q

Quadrupole mass spectrometer   A mass spectrometer that uses a radio frequency electric field between four electrodes to determine which gaseous species with specific charge-to-mass ratio can traverse from the ionizer to the collector. See Mass spectrum.

Quality, laboratory (R&D)   Obtaining data that is reproducible and accurate. Reporting data in a manner such that the results can be reproduced by others.

Quality, product (manufacturing)   The ability of a product to meet the customer's expectations based on cost, appearance, performance, lifetime, reliability, etc. Ability to meet standards.

Quality audit (manufacturing)   An internal assessment of all phases of production that lead to a quality product. Includes considerations such as adherence to MPIs, information feedback, operator morale, consideration of suggestions offered by operators, etc.

Quality control (QC) (manufacturing)   A procedure for monitoring quality and establishing methods for feedback into production.

Quartz (glassware)   An inappropriate term for fused (vitreous) silica. Often called **Fused quartz**. See Vycor™.

Quartz (mineral)   A natural crystalline mineral that undergoes several transformations on heating. Quartz melts and becomes molten silica (SiO_2) at 1713°C.

Quartz crystal monitor (QCM) (deposition rate) Quartz crystal deposition monitors measure the change in resonant frequency as mass (the film) is added to the crystal face.

Quasi-reactive deposition (PVD technology) Deposition of a compound from a compound source where the loss of the more volatile species is compensated by having a partial pressure of reactive gas in the deposition environment. Example: Quasi-reactive sputter deposition of ITO from an ITO sputtering target using a partial pressure of oxygen in the plasma. See Reactive deposition.

R

Rabbit (electroplating)   Trial part sent through the plating line. See Dummying.

Racetrack (sputtering)   The pattern that is eroded by sputtering on a planar magnetron sputtering target.

Rack   Structure to hold parts for processing, such as cleaning or electroplating. See Fixture.

Racking or "to rack"   To mount the parts into a rack or fixture (i.e., "to rack them"). See Un-rack.

Radiant heating (film deposition)   Heating of a surface by radiation from a hot surface. Example: Heating of a substrate from a quartz lamp in vacuum.

Radiation-enhanced diffusion   Enhancement of the diffusion rate by radiation damage from heavy-particle irradiation that generates lattice defects in the near-surface region.

Radiation equation   An equation that provides the intensity of radiation from a hot surface. The radiant energy E from a hot surface is given by $E = \delta T^4 A$ where δ is the emittance of the surface, T is the Kelvin temperature, and A is the area of the emitting surface.

Radiation shield   An optical baffle that is used to contain radiation or prevent radiation from reaching a surface.

Radical   A group of atoms that form an ionic group having one or more charges, either positive or negative. Example: The hydroxyl radical OH^-.

Radio frequency (rf)   An alternating potential (AC) within a certain frequency range. There is no sharp distinction between radio waves and microwaves but typically rf frequencies start at about 50 kHz and extend to 100 MHz, with 13.56 MHz being a common industrial rf frequency. See Audio frequency; Microwave frequency.

Radio frequency (rf) sputtering   Physical sputtering, generally of an electrical insulator, where the high negative electrical potential on the surface is achieved by alternately polarizing the surface positively and negatively at a rate greater than about 50 kHz. During the positive half-cycle, surface charging is neutralized by electrons from the plasma. During the negative half-cycle, ions are accelerated from the plasma to sputter the surface. See AC sputtering.

Rain (vacuum technology)   Vapor phase condensation of water when a chamber with high humidity air is pumped so fast that the gas temperature is lowered below the dew point.

Random arc (plasma)   Cathodic arc where the arc is allowed to move randomly over the cathode surface. See Arc source.

Raoult's Law (evaporation)   Raoult's Law states that constituents of a liquid vaporize at a rate proportional to their vapor pressures.

Rapid thermal chemical vapor deposition (RTCVD)   Chemical vapor deposition using rapid heating and cooling to deposit a coating.

Rapid thermal processing (RTP)   Heating process characterized by rapid heating to a high temperature, a short time-at-temperature, then a rapid cool-down. The heating mostly affects the near-surface region. Example: RTP diffusion into a surface.

Rate-of-rise (vacuum technology)   The time for the pressure in a system to rise a specified amount with no vacuum pumping taking place. Generally the leak-up pressure range is specified, i.e., from 10^{-4} Torr to 10^{-3} Torr. The leak-up rate is an indication of the presence of outgassing, desorption, virtual leaks, and real leaks. Also

called **Leak-up-rate**.

RCA cleaning process (semiconductor processing)   A cleaning procedure widely used for cleaning silicon wafers. Also called a **Modified RCA cleaning process**.

Resputtering rate (ion plating)   The rate of sputtering of the depositing film material due to the concurrent energetic particle bombardment of the growing film. Example: About 20 to 40% resputtering is necessary to completely disrupt the columnar morphology of the depositing film material (Thornton).

Reactant availability (reactive deposition)   The availability and chemical reactivity of the reactive gas over the surface of the film being deposited. Since the surface of the film is continually being buried, reactive gas availability is an important parameter in the reactive deposition process.

Reaction probability (reactive deposition)   The probability that a reactive gas species impinging on a surface will react with the surface to form a compound. The probability depends on the reactivity of the species, residence time on the surface, surface coverage, surface mobility, reaction-enhancing processes such as concurrent electron or ion bombardment, etc.

Reactive deposition (film formation)   Film deposition process in which the deposited species reacts with an ambient gas, an adsorbed species, or a co-deposited species to form a compound material. See Quasi-reactive deposition.

Reactive evaporation (film deposition)   Evaporation in a partial pressure of reactive gas in order to deposit a compound film material. See Reactive deposition.

Reactive ion beam etching (RIBE) (cleaning)   Chemical etching of a surface under bombardment by a reactive ion beam from an ion source that is usually collimated and often monoenergetic.

Reactive ion etching (RIE) (cleaning)   Chemical etching of a surface under bombardment by low-energy reactive ions that are generally accelerated from a plasma of the reactive gas.

Reactive plasma cleaning (cleaning)   Reaction of contaminants with reactive species to form volatile compounds.

Reactive plasma etching (RPE) (cleaning)   Chemical etching of a surface in contact with a plasma of the reactive gas. See Reactive ion etching.

Reactively graded interface (film formation)   A graded interface formed by changing the availability of the reactive gas during the formation of the interfacial region. Example: Grading the film composition from titanium to TiN_{1-x} to TiN by changing the availability of the nitrogen during reactive deposition.

Reactor, cold wall, CVD   Reactor furnace where the CVD gases are heated by the hot substrate and the walls of the containing structure are cold.

Reactor, CVD   The furnace in which the CVD process takes place. See Reinberg reactor.

Reactor, CVD, fluidized bed   A means of floating, stirring, and mixing parts in a heated chamber using a flow of gas containing the chemical vapor precursor. Vibratory action can also be used to aid in moving the parts. Particles can be added to the parts to keep them separated during deposition. See Pack cementation.

Reactor, CVD, hot wall   Reactor furnace where the CVD gases and the substrates are heated by conduction and radiation from the containing structure (furnace).

Reactor, CVD, Reinberg   A parallel-plate, rf-driven reactor for plasma enhanced CVD (PECVD).

Real gas   A gas that does not obey the Ideal Gas Law because of molecule-to-molecule chemical interactions. Example: Water vapor at room temperature.

Real surface (substrate)   The substrate surface that must be processed in film deposition. The real surface often has reaction layers, such as oxides, contaminant layers, such as adsorbed hydrocarbons, and some degree of particulate contamination. Also called **Technological surface**.

Receptor films (heating)   Thin metal films used for microwave heating of packaging materials.

Recoil implantation (cleaning, film formation)   When a high-energy bombarding species imparts enough energy to a surface atom to cause it to be recoil implanted into the lattice as an interstitial atom.

Recombination (plasma chemistry)   The combining of a positive ion with an electron so as to form an uncharged species. This process mostly occurs on surfaces and the process gives up the ionization energy to the surface and the neutral species.

Recommended practice   A type of specification that has not gone through the rigorous review procedure as that of a Standard. Example: AVS Recommended Practices for calibrating pump speed. See Standard.

Recontamination (cleaning)   The contamination of a cleaned surface. Recontamination depends on the chemical reactivity of the surface, the environment, and the exposure time.

Recrystallization   Change of phase or crystal growth orientation in a material due to temperature or stress. Example: Devitrification of glass.

Redeposition   When a material that has been vaporized deposits on the surface from whence it came. Example: Backscattering in a gaseous environment.

Reducing agent (cleaning)   A material that adds electrons and elemental species such as hydrogen to a compound, often forming a volatile species. Example: Hydrogen reduction of the oxide on a metal surface by dry

hydrogen gas to form water and an oxide-free metal surface.

Reduction reaction   A chemical reaction in which a compound gains an electron. Also: The addition of hydrogen or the loss of oxygen.

Reduction reaction (CVD)   Reduction of a chemical vapor precursor to obtain a condensable film material. Example: $TiCl_4 — 2H_2 —> Ti + 4HCl$.

Reflected high-energy neutrals (sputtering)   In the sputtering process, a portion of the high-energy bombarding ions becomes neutralized and is reflected from the cathode (target) surface. If the gas pressure is low, these high-energy particles are not thermalized and bombard the growing sputter-deposited film, and influence film properties such as residual film stress.

Reflected power, rf (plasma technology)   rf power that returns to the power supply because of poor impedance matching between the load and the power supply. Reflected power should be minimized by proper impedance matching.

Reflow (surface)   Heating a surface to melt and flow the surface.

Refraction   The bending of light as it passes from one medium to another because of the change in the velocity of the light in passing from one medium to the other.

Refractive index (optics)   The ratio of the velocity of light in vacuum to the velocity of light in a material. Also the sine of the angle-of-incidence of the light beam in vacuum to the sine of the angle-of-refraction of the light as it enters the second media.

Refractory material   A material that has a very high melting point.

Refrigerator effect (cleaning)   When an item is in the back of the storage area and stays there much longer than other items. Opposite of the first-in-first-out storage procedure.

Regeneration (vacuum technology)   Warming up a cryosorbing material to cause the adsorbed gases to be volatilized. Regeneration may be to room temperature (activated carbon) or to higher temperatures (Zeolites).

Regeneration cycle time   The time necessary to regenerate the cryosorbing material and to return it to its operating temperature.

Reinberg reactor (PECVD)   A parallel-plate, rf-driven reactor for plasma enhanced CVD (PECVD). See Reactor, CVD.

Relative humidity   The ratio of the amount of water vapor in a gas to the amount it could hold at saturation expressed as a percent. See Humidity.

Release layer (vacuum technology, PVD technology, electroplating)   A layer of material (**Release agent**) that ensures poor adhesion between the deposited film and a surface. Used in cleaning excess material from vacuum surfaces and to release a deposit from a mandrel to become a freestanding structure.

Remote plasma source (plasma)   A plasma source where the plasma is generated in one region and used in another (downstream) region.

Remote region (plasma)   The **Afterglow** or **Downstream region**.

Removable surfaces (vacuum technology)   Surfaces, such as fixtures, that are routinely removed from the system or surfaces such as liners that can be removed from the system for cleaning. See Non-removable surfaces.

Repeatability (manufacturability)   The ability to obtain the same results on a number of trials or measurements. See Precision.

Reproducibility   When the process and/or product can be duplicated from run-to-run within specified tolerances.

Residence time (vacuum technology, film formation)   The amount of time that an impinging atom or molecule spends on a surface before it leaves the surface.

Residual film stress (film formation)   The residual compressive or tensile stress in a film that results from the growth process (**growth stress**), phase change, and/or differences in the coefficient of thermal expansion of the film and substrate (**thermal stress**). Not a function of film thickness. Can vary through the thickness of the film and be anisotropic with direction in the film. See Total film stress.

Residual gas (vacuum technology)   The gases in the vacuum system at any specific time during pumpdown or processing.

Residual gas analyzer (RGA)   Device for measuring the species and the amount of residual gases in a vacuum system. See Mass spectrometer; Partial pressure analyzer.

Residue (cleaning)   Any undesirable material from the chemicals used in processing that remains on a surface after a processing step.

Resistance heating (evaporation)   The Joule or I^2R heating of an electrical current (I) passing through a material having an electrical resistance (R).

Resistivity   See Sheet resistivity (thin film), Specific resistivity (bulk).

Resistivity of water (cleaning)   The electrical conductivity of water as measured between probes spaced one centimeter apart. Example: 18 megohm-cm. One measure of the purity of the water. See De-ionized water; Ultrapure water; Hard water; Soft water.

Reverse engineering   The process of taking a completed structure and determining the structure, the materials, and the techniques used to build the structure.

Reverse osmosis (water purification)   Using high pressure (150 to 600 psi) to force water through a membrane that will not pass ions such as sodium, iron, manganese, calcium, etc.

Rework   To take a part that has been rejected in inspection and repair or redo the reason for the rejection.

rf plasma source   A plasma source that uses radio-frequency radiation to excite the plasma. The design may use a coupled plasma such as a parallel plate design, or an inductively coupled plasma using a coil design. See Plasma source.

Rinse (cleaning)   To remove residual processing chemicals with a material that has no detrimental residue. Example: Rinsing with ultrapure water. See Drag-out.

Rinse aid (cleaning)   Chemical that reacts with a polymer surface (which may have been originally hydrophobic) to make the surface hydrophilic—that is, to make water sheet over the surface more easily and not ball up on the surface.

Rinse fluid (cleaning)   A fluid used to displace fluids that have potential residue materials. Examples: Pure water; Ultrapure water; Perfluoro-N-methyl morpholine (3M PF-5052 "spot-free" rinse agent).

Rinse-to-resistivity (cleaning)   Rinsing a surface in pure water until the water retains a specific resistivity such as 10 megohm-centimeters.

Rinsing, cascade (cleaning)   Rinsing using a series of containers (**Rinse tanks**) having increasingly pure water. Water generally flows over the lip of one container into the next container having a lower purity water. The surface being rinsed goes from the lower purity to the higher purity rinse tank.

Robust process   A process that has wide parameter windows.

Roentgen   The amount of X-ray or gamma radiation that will produce 1 esu of charge of either sign when passing through one (1) cubic centimeter of air at 760 Torr and 0°C.

Roll coater   See Web coater.

Root mean square   The square root of the average value of the squares of the values measured.

Roots blower (vacuum technology)   A compression-type mechanical pump that uses lobe-shaped interlocking rotors to capture and compress the gas. The Roots pump uses tight mechanical tolerances for sealing (no oil) and so is sometimes classed as a dry pump. See Vacuum pump.

Rotary vane pump   A displacement pump where the compression occurs in a non-symmetric chamber being swept by a rotor having an oil-sealed sliding-vane. See Vacuum pump.

Rotatable cylindrical magnetron (sputtering)   A water-cooled tubular sputtering target containing a magnetron magnetic field arrangement such that the wall of the tube is rotated through the magnetic field, producing uniform sputter-erosion of the whole surface of the tube. See Magnetron.

Rottenstone (abrasive) A solid block of abrasive that continuously wears during abrasion.

Rough vacuum (vacuum technology)   Pressure from atmospheric to about 50 mTorr.

Rough vacuum (vacuum technology)   Pressure from atmospheric pressure to the crossover pressure. See Crossover pressure.

Roughing pump (vacuum technology)   Vacuum pump used to lower the pressure in the system through the rough vacuum range. The roughing pump is often also used as the backing pump for a high-vacuum pump. See Backing pump; Vacuum pump.

Roughness, surface (R$_a$) The arithmetic mean of the departure of the roughness profile from a mean value. The R$_a$ is also called the **Center line average (CLA)**.

Round-robin (test)   Series of procedures, measurements, or processes performed by different groups for comparison before the procedure, measurement technique, or process is incorporated into a Standard. See Standard.

Rugate filter (optics)   A film in which the refractive index varies continuously and periodically with the coating thickness.

Run, deposition   Each deposition process including pumpdown-deposition-letup to atmosphere. See Cycle (process).

Rust   Visible corrosion product on ferrous alloys. Usually friable.

Rutherford backscattering spectrometry (RBS) (characterization)   A non-destructive technique for depth profiling the chemical composition of a material to a depth of several microns. The probing species is a high-energy (MeV) light (He$^+$) ion and the detected species are energy-analyzed helium atoms that have been backscattered from the atoms in the solid.

Sacrificial protection (corrosion)   A form of corrosion protection where one material corrodes in preference to another, thereby protecting it. Examples: Zinc and cadmium on steel; Aluminum on steel.

Sampling method, statistical (manufacturing)   The method used for selecting sample(s) that, when characterized, will be representative of the batch as a whole or for establishing position equivalency on a fixture. Sampling can vary from 100% (such as tape-testing 100% of a mirror surface) to periodic sampling. Sampling is used to characterize the product during the manufacturing process.

Sanitary pipe (vacuum technology)   Elastomer-sealing plastic or glass components used in the food industry that are suitable for use in vacuum technology for some applications, such as assembling exhaust manifolds.

Saponification (cleaning)   The conversion of oils into soaps by **Alkaline hydrolysis**.

Sapphire (substrate)   Single-crystal or gem-quality aluminum oxide (Al$_2$O$_3$). See Corundum.

Saturation vapor pressure   The maximum pressure that can be exerted by a vapor in thermodynamic equilibrium with a surface of the material. Example: The saturation vapor pressure of water vapor at room temperature is about 20 Torr. Also called **Equilibrium vapor pressure**. See Supersaturation.

Scale (cleaning)   A thick layer of oxide that forms on some metals during high-temperature processing. Example: Mill-scale on steel directly from the steel mill.

Scale-up (manufacturing)   The ability to increase product throughput to the desired level using proven processes by decreasing the cycle-time, building larger equipment, increasing the operating time, etc. See Manufacturability.

Scanning auger microscopy (SAM) (characterization)   A scanning surface analytical technique that uses an electron beam as the sampling probe and Auger electrons as the detected species to give the composition of the surface. See Auger electron spectroscopy (AES).

Scanning electron microscopy (SEM) (characterization)   The SEM uses the secondary electrons from an electron-bombarded surface to form an image of the surface morphology. The magnification can be varied from several hundred diameters to 250,000 diameters with high lateral and vertical resolution.

Scanning laser acoustic microscopy (SLAM) (characterization)   In SLAM a pulsed laser introduces a thermal wave into the material. A discontinuity in the material through which the thermal pulse passes can give rise to acoustic emission, which is then detected.

Scanning thermal microscopy (SThM) (characterization)   An AFM that uses a thermocouple junction as the probe tip, which can detect variation in temperature over a surface to a lateral resolution of about 10 nm.

Scanning transmission electron microscopy (STEM) (characterization)   The STEM uses the transmission of electrons through a thin film to image the microstructure of the film to a resolution of several Ångstroms.

Scanning tunneling microscopy (STM) (characterization)   The STM measures the electrons that tunnel between a probe tip and a surface. The system is typically operated in a constant current mode and the movement of the tip is determined to about 0.1 Å.

Scatterometry (characterization)   Scatterometry measures the angle-resolved scattering of a small spot of laser-light incident on a surface. The distribution of the scattered energy is determined by the surface roughness.

Scoring (surface)   The formation of a severe scratch or cut on a film or surface. Often used to provide a source of fracture for breaking brittle materials or pulling a film from the surface.

Screen   A **Sieve** having a screen with a specific opening size to allow classification of particles as to their size. Usually used as a series of screen sizes. See Mesh sizing.

Scrubbers (vacuum technology, CVD)   Units placed in the exhaust side of a pumping system to remove particulates and toxic gases. Generally the scrubbers use water to collect particles and chemicals, though in some cases the gases are burned to form solids. Example: SiH_4 burned to form SiO_2.

Sculpted thin films   Films grown with the columnar growth controlled by varying the angle-of-incidence to give various shapes to the columns.

Scum (cleaning)   Layer of contamination that floats on the surface of a liquid. Scum can be removed mechanically (**Skimming**) or by using overflow tanks.

Scum (evaporation)   Material that is on the surface of molten material and that is visually obvious.

Seal, bakeable (vacuum technology)   A seal that can be heated to an elevated temperature, typically 400°C.

Seal, elastomer (vacuum technology)   A seal using an elastomer to provide the deformation and restoring force needed to form a vacuum-tight joint.

Seal, demountable (vacuum technology)   A seal designed to be disassembled and reassembled easily using a gasket. The sealing gasket may be reusable or replaced each time the seal is disassembled. Also called a **Nonpermanent seal.**

Seal, metal, shear (vacuum technology)   A seal formed by having a knife-edge shear into a soft metal.

Seal, metal, spring (vacuum technology)   A metal seal having a crossection in the form of a "C" containing a canted coil spring that causes an expansion force to open up the "C."

Seal, permanent (vacuum technology)   A seal that is designed so as not to be easily disassembled. Example: A weld or braze joint.

Seal, spring-loaded (vacuum technology)   Non-elastomeric polymer seal where the restoring force is supplied by an internal metal spring.

Sealant   Material used to plug a leak.

Sealing surface (vacuum technology)   The smooth surface to which an elastomer gasket deforms and seals.

Second surface (optical)   The surface of the optical substrate opposite the incoming radiation. Example: Second surface mirror that is metallized on the "backside" of the glass. See First surface.

Second surface coating (decorative coating)   The reflective coating (usually aluminum) that is used underneath the lacquer coating. The lacquer coating (topcoat) is used to give color and texture to the coated part.

Secondary electron emission   The emission of electrons under electron or ion bombardment.

Secondary ion mass spectrometry (SIMS) (characterization)   A surface analytical technique that uses high-energy ions as the probing species and sputtered ions from the surface as the detected species.

Secondary standard   A standard that is commonly used to calibrate components that are in use. The secondary standard is periodically checked against a primary standard at the manufacturing site. See Primary standard.

Seed (crystal growth)   Single-crystal particle (**Seed-crystal**) that acts to nucleate growth of a single-crystal ingot.

Seed (film formation)   Defect in a deposited film due to particulate contamination of the growing film during deposition.

Seed (glass)   Defect in glass due to a foreign particle.

Seed layer (film formation)   A layer, often close to one monolayer thick, that acts as a nucleating layer for subsequent deposition.

Seizing (mechanical)   The stopping of moving parts in contact by virtue of galling, deformation, and adhesion.

Selected area diffraction (SAD)   Electron diffraction done on selected areas of a film in a Transmission electron microscope (TEM) to determine crystal structure.

Selective deposition   Deposition on a local area. May be due to masking, local areas of heating, nucleation sites, or local application of electrolyte solutions (brush plating in electroplating).

Self-bias (plasma technology)   An electrical potential on a surface generated by the accumulation of excess electrons (negative self-bias) or positive ions (positive self-bias). See Sheath potential.

Self-ion (sputtering, sputter deposition)   An ion of the sputtered target material that can bombard the target, giving **Self-sputtering**. See Film ion.

Self-sputtering   Sputtering by an ion of the target material being sputtered (**Self-ion**). See Film ion.

Semi-aqueous cleaning (cleaning)   Where a non-aqueous material is used for cleaning but water is used in some stage of the cleaning process. Example: A mixture of a terpene with a surfactant for cleaning and a water rinse to remove residue-producing material. See Aqueous cleaning; Non-aqueous cleaning.

Semiconductor grade (cleaning)   Materials that meet the purity specifications set by the semiconductor industry.

Semiconductor material A material whose electrical conductivity is intermediate between a good conductor and an insulator. The resistivity is generally strongly temperature-dependent and can be varied by doping. See Dopant.

Sensitivity (sensor) The response of a sensor to a small change in the condition being measured. See Sensor.

Sensitization (surface) The production of unsatisfied chemical bonds on a surface that increases the chemical reactivity of the surface. Often sensitization is a temporary condition so the **Time-to-use** must be specified.

Sensor (vacuum technology) A device that detects a property or a condition of a system. The output of a sensor can be used by a microprocessor to control the system. Examples: Vacuum gauge; Temperature gauge; Flow meter. See Feedback.

Sequestering agents (cleaning) Materials that react with the metal ions in hard water, keeping them in solution, thus preventing them from reacting with cleaning agents and forming insoluble precipitates. These materials can present pollution problems if used in large quantities. Examples: Orthophosphates; Orthosilicates.

Serial co-sputtering (PVD technology) When material from one sputtering target is deposited onto another sputtering target from which it is sputtered to produce a graded or mixed composition.

Set The permanent or semi-permanent shape that a polymer assumes under a load that relieves the elastic stress in the material. A material, such as Teflon™, that "takes a set" is not a good material for an elastomer seal.

Shall Term used in a specification or a manufacturing process instruction (MPI) that indicates a mandatory procedure. Example: The gloves shall be discarded after each use. See May; Should.

Shaped anodes (electroplating) Anodes that are shaped (often conformal to the cathodic substrate) to produce a uniform field between the anode and the cathode and to reduce high field regions on the cathode.

Shard Small fragment of a brittle material. Examples: Glass shards in glass bead blasting; Glass shards formed when tempered glass is fractured.

Shear stress (adhesion) Stress parallel to an interface. See Tensile stress; Compressive stress.

Sheath (plasma) The region near a surface whose properties are affected by the bias on the surface. Examples: Wall-sheath; Anode-sheath.

Sheath potential (plasma) The potential across a sheath. Example: The potential across the wall sheath is typically a few eV, with the plasma being positive with respect to the wall due to the higher mobility of the electrons as compared to the ions.

Sheet resistivity The resistance from side-to-side of a square area of any size on a film expressed in **Ohms-per-square**. To obtain the specific resistivity (ohm-cm) of the coating material, the film thickness must be known. See Electrical resistance.

Sheeting (cleaning) The uniform flow of a fluid over a surface. If the sheeting is not uniform then contamination is suspected. See Legs.

Sheeting agent (cleaning) A material applied to a surface to cause water to flow (**Sheet**) evenly from the surface. This helps to reduce residues (e.g., **Water spots**) left on the surface. A common sheeting agent is paraffin in a solvent.

Shelf samples Samples that are placed in a normal environment to age normally and be available for comparison in the future. Also called **Archival samples** or **Control samples**.

Sherardizing Coating with zinc by mechanically tumbling a part in hot zinc powder. See Mechanical plating; Peen plating.

Short-term exposure limits (STEL) (safety) The short-term (15 minutes) exposure limits to hazardous materials as established by OSHA. See Permissible exposure limits (PEL); Time-weighted average (TWA).

Shot peening (substrate) Mechanically workhardening a ductile surface by repeatedly striking it with hard balls, usually entrained in a high-velocity gas stream.

Shot peening (postdeposition processing) Densifying a ductile film by repeatedly striking it with hard balls, usually entrained in a high-velocity gas stream. Peening compacts the film and closes porosity.

Should Term used in a specification or an MPI that indicates a good practice but which is not mandatory. Example: Gloves should be discarded after use. See Shall; May.

Shrinkage (sintering) The reduction in volume due to firing.

Shutdown (vacuum technology) Putting equipment in a safe and non-contaminating condition in preparation for non-use. Shutdown of a vacuum system may mean turning it off or may mean leaving the system under active high-vacuum pumping.

Shutter (vaporization) A movable optical baffle between the vaporization source and the substrate that prevents contaminants from the source from depositing on the substrate during the initial heating of the source. The shutter also minimizes radiant heating of the substrate before vaporization begins. The shutter can also be used to establish the deposition time.

Shutter, window A movable optical baffle in front of a window to minimize the deposition of film material on the window when it is not being used.

Siemens A unit of conductance equal to the reciprocal

of the resistance in ohms. See Mho.

Silica (substrate)　Silicon dioxide (SiO_2). Usually in the form of a glass called **Fused silica** or **Fused quartz**. The crystalline material called **Quartz**.

Silica, fumed (polishing)　Silicon oxide abrasive prepared by oxidizing vaporized silane in the gas phase.

Silica gel (desiccant)　A desiccant material composed of amorphous silicon oxide that is porous. The pore sizes range from 1.4 nm to 3.0 nm. See Molecular sieve.

Silicon carbide (abrasive)　Silicon carbide (SiC) abrasive.

Silicone oil (vacuum technology)　A heavy, low-vapor-pressure silicone-based (rather than hydrocarbon-based) oil that is commonly used in diffusion pumps and is sometimes used as a lubricant in vacuum systems. In diffusion pumps silicon oils are preferable to hydrocarbon oils since they are less prone to oxidation.

Silvering (chemical solution)　The deposition of silver from a solution by a catalyzed reduction reaction on the surface. Used to coat surfaces for mirrors and vacuum insulation. Example: Vacuum-insulated flasks (**Dewar flasks**).

Single-unit processing (PVD technology)　Processing one (or a small number of) units at a time in contrast to processing a number of units each cycle (Batch coating). Example: Processing compact discs one at a time with a cycle time of less than 3 seconds.

Sintering　To bond particles together by solid state diffusion to the contact points at an elevated temperature and sometimes under pressure. In many cases a small amount of bonding fluid may be present such as in glass-bonded "sintered" alumina. See Hot isostatic pressing (HIP).

Sizing (cleaning)　The lubricant applied to a thread to aid in weaving it into cloth. The sizing agent is often polyethylene glycol, which is water soluble and can be removed by multiple washing. Sodium silicate also is used as a sizing agent but it is difficult to remove by washing.

Skim (cleaning)　To mechanically remove material that is floating on top of a fluid. Example: Oil on water. See Oleophillic filters.

Skin (sintered material, sputtering target)　The dense surface layer that is sometimes formed on sintered materials.

Skull (evaporation)　The solid liner that forms between a molten material and a surface. The skull may be due to cooling (such as a molten material in contact with a water-cooled copper hearth) or may be due to the formation of a reaction layer (such as molten titanium in contact with a carbon liner, giving a TiC skull).

Slip agents (web coating)　Agents added to polymer films to increase the friction of the surface. Slip agents may be inorganic particles added to the film material or may involve chemical surface treatment.

Slip-cast　A suspension of particles (the **Slip**) that is formed into a shape, such as a plate or a ribbon, before solidification. The solidified slip is then fired to drive off volatile materials and bond the particles together by fusion and/or sintering. Example: Slip-cast alumina.

Slitting (web)　Cutting the web in the machine direction to create a narrow web.

Slurry polishing　Polishing of a surface by particles in a fluid suspension (slurry) passing over a surface. If the slurry is very dilute the polishing may be called **Water polishing**.

Smock (cleaning)　Lightweight collared coat, usually with a front opening and long sleeves, used to protect clothing and/or contain particles produced by clothing and skin. Often made with an impervious material, such as Tyvek™, a close-woven cloth from a long-fiber thread, such as Nylon™, or a moisture-breathable fabric, such as GoreTex™. Also called a **Lab coat**.

Smut (cleaning)　Residue of very fine particles on a surface after chemical etching or preferential sputtering. The particles are of second-phase material that are not attacked by the etchant. Example: Copper smut left after etching an Al-2%Cu alloy with NaOH.

Snell's Law　The index of refraction of a material is the ratio of the sine of the angle of incidence of the radiation on a surface (from vacuum) to the sine of the angle of refraction in the material. See Index of refraction.

Snow (cleaning)　Solid material formed from a gas or a fluid, usually by expansion and cooling (e.g., CO_2), used to clean a surface.

Soak (cleaning)　To leave in a fluid for a long period of time.

Soak (heating)　To leave at a high temperature for a long period of time.

Soak cleaning　See Immersion cleaning.

Soap (cleaning)　The water-soluble reaction product of a fatty acid ester and an alkali, usually sodium hydroxide. Used to emulsify oil contaminants.

Soft wall clean area (cleaning)　A clean area defined by hanging PVC plastic drapes where the filtered air flows from the ceiling downward and out under the drapes. The drapes may be in the form of strips (**Strip curtains**).

Soft water (cleaning)　Water that is free of ions, such as calcium and magnesium, that can form water-insoluble precipitates and residues. Soft water is produced by exchanging the ions with sodium and chlorine ions from NaCl. Sometimes used in rinsing before the final rinse, which should be done using pure or ultrapure water. See Water.

Sol gel coating　The coating of a surface with a fluid

sol, which is a stable suspension of colloidal particles. The sol is then converted into a rigid porous mass called a gel, which is heated to melt and sinter the mass into a solid thin film.

Solar control coating A type of thermal control window coating that consists of a thin film structure designed to reflect incident solar radiation and prevent it from heating the interior of a room while retaining reasonable transmission in the visible radiation spectrum. See Low-E coating.

Solder alloy A metallic material that melts at a temperature less than 450°C and is used to join two materials together. See Solder, tin-lead; Braze alloy.

Solder, tin-lead (vacuum technology) A solder alloy that contains tin and lead (63/37, 60/40) and does not contain any volatile constituents such as zinc or cadmium. It is thus suitable for use in a vacuum system.

Solid lubricant (vacuum technology) A non-liquid material that provides lubrication and does not creep away from the point of application the way a liquid lubricant does.

Solid lubricant, low-shear metals A solid lubricant used in high-torque application where lubrication is provided by deformation and shear of a non-workhardening metal. Examples: Silver; Lead.

Solid lubricants, low-shear compounds A solid lubricant used in low-torque applications where the lubrication is provided by shear between crystallographic planes. Example: MoS_2.

Solids content The amount of solid material left after the solvents have been volatilized. An important property of material deposited by flow coating, such as basecoat material.

Solidus (phase diagram) The lower boundary of the liquid + solid region of the phase diagram. See Liquidus.

Solubility parameter (cleaning) The amount of a specific material that a unit volume of a solvent will take into solution. Used to compare the relative cleaning power of cleaning solutions.

Solute The material that goes into solution.

Solution, chemical A homogeneous mixture of two or more chemicals, the composition of which can be varied within limits.

Solvent (cleaning) A material capable of dissolving or taking into solution another material (the solute).

Sonoluminesce (cleaning) The ultrashort bursts of light emitted by bubbles collapsing in a fluid.

Soot (CVD, reactive deposition) Ultrafine particles formed by gas phase decomposition (CVD) and nucleation. See Ultrafine particles.

Sorption The taking up of a gas by a solid or liquid material (sorbant), either by adsorption or absorption.

Sorption pump Vacuum pump that operates by sorption of gases and vapors on surfaces that are usually cold. See Vacuum pump.

Sour cleaning bath (cleaning) A chlorinated solvent bath that has become acidic by reaction with water to form HCl.

Space charge The net charge in a volume of space caused by an excess of one charged species over another. Example: An excess of electrons and negative ions over positive ions will result in a negative space charge.

Spare parts (vacuum technology) Spare parts to replace parts that, if they fail, will prevent use of the equipment. Also called **Operational spares** (preferred). Examples: Spare roughing pump; Spare O-rings.

Sparger (cleaning, electroplating) Perforated pipe distributor for fluids or gases used in the bottom of fluid tanks for agitation.

Spark discharge plating The transfer of material from a cathodic electrode to the anodic substrate in a periodic low-voltage, high-current arc in air or an inert gas.

Specific cleaning (cleaning) Cleaning procedures directed toward removing specific contaminants. Example: Removal of hydrocarbon contaminants by oxidation. See Gross cleaning.

Specific gravity (sg) (cleaning) The ratio of the density of a material to the density of water, at a specific temperature.

Specific gravity (solution strength) A method of specifying solution strength. Example: Sulfuric acid varies from an sg of 1.0051 for a 1% aqueous solution to 1.8305 for a 100% saturated solution.

Specific heat The quantity of heat needed to raise the temperature of a unit amount of material one degree.

Specific volume The volume per unit mass of a material.

Specification, process The formal document that contains the "recipe" for a process and that defines the materials to be used, how the process is to be performed, the parameter windows, and other important information related to safety, etc. Information on all critical aspects on the **Process flow sheet** should be covered by specifications. See Process flow sheet.

Spectrophotometer An instrument that measures radiation intensity at a specific frequency and over a broad band of frequencies.

Specular reflection, optical Reflection at a specific angle determined by the angle-of-incidence of the incident beam. See Diffuse reflection.

Speed The rate of change of position. Speed is a **Scalar** quantity. Examples: Miles-per-hour; Feet-per-second. See Velocity.

Speed, pump The volumetric rate of gas flow through

a pump as measured in liters per second, ft³/min, m³/hr, etc. In order to calculate the mass flow rate (Torr-liters per second), the pressure must be specified.

Sphygmomanometer (pressure)   A meter for measuring blood pressure to several hundred millimeters of mercury.

Spin coating (semiconductor manufacturing)   Coating of a rapidly rotating surface with a fluid by applying the fluid at the center of the axis of rotation, and letting centrifugal force carry the fluid to the edges where the excess is flung-off.

Spin dry (cleaning)   Removing most of the fluid from a surface by spinning at a high rate so that centrifugal force carries the fluid to the edge where most of the fluid is flung-off.

Spinning rotor gauge (vacuum technology)   A type of **Viscosity vacuum gauge** that measures the deceleration of a levitated ball due to frictional drag with the gases present. Gauge output depends on the density and composition of the gases present.

Spit (evaporation)   A molten droplet of the evaporant ejected from the molten surface. Spits generally result from vapor bubbles rising through the molten material. See Boiling beads.

Splat cooling (thermal spray coating)   The rapid cooling of a molten droplet of material.

Split flow (leak detection)   When part of the helium flow passes through the leak detector and part through the high-vacuum pumping system. See Full flow.

Sport (statistics)   Data point, event, or product that occurs outside the norm for no obvious reason. Often disregarded in statistical analysis.

Spot cleaning (cleaning)   Cleaning of a localized area on the substrate.

Spray (cleaning, rinsing)   Spraying (in air) with an agent such as a solvent at a low pressure (100 psi) or a high pressure (1000 psi). Note: Some people use the term spraying to describe the use of high-velocity fluid jets in the fluid of a cleaning tank. The author would call this **Fluid jet agitation**.

Spray rinsing (cleaning)   Spraying with soft, pure, or ultrapure water to rinse the surface.

Sputter (PVD)   Ejection of particles from a surface by momentum transfer from an energetic atomic-sized particle impinging on the surface.

Sputter cleaning (cleaning)   Removal of surface material in the deposition chamber by physical sputtering. See *In situ* cleaning.

Sputter deposition (PVD)   A physical vapor deposition process in which the source of the depositing atoms is a surface (target) being sputtered.

Sputter etching (semiconductor processing)   Removal

of material in a specific pattern by sputtering. See **Ion milling.**

Sputter texturing   Surface roughening by preferential sputtering of crystallographic planes or due to isolated inclusions or patches of low-sputtering-yield material on the surface. See Cone formation.

Sputter-ion pump   A capture (getter) pump in which the gettering material is continuously being renewed by sputter deposition. See Vacuum pump.

Sputtered (PVD)   Deposition of material by sputtering (as in "sputtered films"). See **Sputter deposition** (preferred).

Sputtering, chemical   The vaporization of surface atoms by chemical reaction with a reactive bombarding species, resulting in an easily volatilized compound species. Example: Sputter etching of silicon using bombardment with chlorine ions. See Reactive plasma etching (RPE); Reactive ion etching (RIE).

Sputtering, collimated   A sputtering arrangement where the off-normal portion of the flux of sputtered species is eliminated, usually mechanically. See Directed sputtering.

Sputtering, diode   A sputtering arrangement where the cathode electrode is the **Target** that ejects electrons by **Secondary electron emission** and atoms by **Sputtering**. The second electrode is the anode. See Sputtering, triode.

Sputtering, directed   Sputter deposition where the sputtered species form a low-divergence beam before impinging on the substrate. The divergence can be minimized by mechanical means (collimated), ionization, and acceleration, or by "**long-throw**" **sputtering**.

Sputtering, dual cathode (PVD)   When two sputtering targets are electrically connected with each other such that when one target is the cathode, the other is the anode, with the polarity switching at a frequency of less than 50 kHz so each target is acting in a DC diode mode. This arrangement reduces the problems of the "**Disappearing anode effect**" when reactively depositing insulating film.

Sputtering, physical   The physical ejection (vaporization) of a surface atom by momentum transfer in the near-surface region by means of a collision cascade resulting from bombardment by an energetic atomic-sized particle.

Sputtering, pulsed power   A diode configuration in which the negative potential is applied as a fast risetime DC pulse with a zero or reverse potential for a short portion of each cycle in the **Midfrequency AC** range. This minimizes charge buildup and surface flashover (arcing) on a target that is being "poisoned" by a dielectric coating.

Sputtering, self   Sputtering of metals using ions of the same material either originating from the sputtering target or from another source.

Sputtering, triode A sputtering arrangement that has three electrodes. One cathode is the cathode of the gas discharge, another electrode is the anode of the gas discharge (and is often a hot filament electron emitter). The third electrode is a sputtering cathode that provides the sputtered species to be deposited.

Sputtering configuration The geometry used for sputtering. See Magnetron; Deposition systems; Fixturing.

Sputtering efficiency (energy) The amount of energy that is represented by the ejected sputtered atom (vaporization energy plus kinetic energy) to the amount of energy put into the surface by the bombarding species. Sputtering has a very low energy efficiency compared to thermal evaporation.

Sputtering target (PVD technology) The material to be sputtered. Generally a cathodic surface in a gas discharge. See Target.

Sputtering threshold The minimum incident particle energy necessary to cause sputtering.

Sputtering yield The ratio of the number of atoms ejected to the number of high-energy incident ions in the sputtering process.

Stability (film property) The ability of a film to retain a specific value of a property over time, environmental exposure, testing, use, or other condition.

Stability (sensor) The precision or accuracy of the output of a sensor (and/or its associated electronics) over a period of time or under repeated operation.

Stabilizers (cleaning) Materials added to chemicals such as solvents and oxidants to reduce the decomposition rate.

Stack (optics) A multilayer film structure used to obtain a desired property or properties.

Staging ratio (vacuum technology) Ratio of the pumping speed of one pump (or stage) to the next pump (or stage) in a multistage pump or train of pumps.

Stainless steel, austenitic (vacuum technology) A non-magnetic, non-dispersion-hardenable stainless steel composed mainly of austenite (gamma iron with carbon in solution), stabilized by nickel. See Stainless steel, martinsitic.

Stainless steel, low-carbon (vacuum technology) A type of stainless steel having a low carbon content, used in situations where welding can cause precipitation of a carbide phase that can result in galvanic corrosion problems. Example: 304L and 316L stainless steel where the L designates a low-carbon content of 0.035 % or less.

Stainless steel, martinsitic (vacuum technology) A magnetic, dispersion-hardenable stainless steel mostly composed of martensite. See Stainless steel, austenitic.

Standard atmosphere Atmospheric conditions of 760 Torr pressure and 0°C temperature.

Standard conditions See Standard temperature and pressure (STP) conditions.

Standard cubic centimeters per minute (sccm) A gas flow in units of cubic centimeters per minute under standard conditions of temperature and pressure.

Standard temperature (SEMI Standards) Sometimes means room temperature i.e., 21°C ± 6°C (70°F ± 10°F). **DO NOT USE THIS TERMINOLOGY!**

Standard temperature and pressure (STP) conditions (vacuum technology) Conditions of 760 Torr and 0°C.

Static dissipative material Electrically conductive material that prevents static charge buildup. Examples: Electrically conductive gloves; Conductive containers.

Static electricity (cleaning) The electric charge that is built up on an insulator surface, typically by friction and the charge separation associated with the friction. The amount of charge buildup depends on the conductivity of the surfaces and the humidity. Static charge buildup can be a problem with blow-drying insulating surfaces with un-ionized air.

Static fatigue (adhesion) The progressive loss of strength of a brittle material under tensile stress due to the weakening of the crack tip by water molecules.

Statistical design (experiments) A technique of optimizing the information that is obtained from the least number of experiments. Useful for establishing process parameter limits. Also called **Factorial design**. See Parameter windows.

Statistical process control (SPC) (manufacturing) A method of measuring the variations in a processing step to help identify the cause of the variations.

Steam-jet pump (vacuum technology) A kinetic vacuum pump where the gases are entrained in a jet of steam. Useful when there is a lot of particulate matter in the gas to be pumped. See Water jet pump.

Steered arc (plasma technology) A cathodic arc where the arc is moved over the surface under the influence of a magnetic field. See Random arc.

Sterling (silver) Silver with a purity of 0.925 fineness.

Sticking coefficient (film formation) The ratio of the particles that remain on the surface to those striking the surface. Also called **Sticking probability.**

Stitching, interfacial (adhesion) Ion implantation through the interface to improve adhesion by imparting energy to the atoms in the interfacial region by collision.

Stoddard solvent (cleaning) An organic-based solvent that has a low (100°F) flash point, dries slowly, and is a health hazard since it contains benzene. Also called **Mineral spirits; Petroleum solvent; Naphtha distilled solvent.**

Stoichiometric compound   A compound material that has the correct atomic ratios for all lattice sites to be occupied for the specific phase of the material. Examples: CuO (1 : 1) or Cu_2O (2 : 1). See Sub-stoichiometric.

Stoichiometry   The numerical ratio of atoms in a compound.

Storage, active (cleaning)   Storage in an environment where contaminants are continually being removed. Example: An ultraviolet-ozone cabinet where hydrocarbons are continually being oxidized. See Storage, passive.

Storage, passive (cleaning)   Storage in an environment that has been cleaned but is not being cleaned while the substrate is in the storage environment. Example: Cleaned glass container. See Active storage.

Stones (glass)   Second-phase inclusions in the glass that produce visually observable defects. See Seeds.

Strain point (glass)   The temperature above which atoms and molecules will move so as to relieve any stress present in the glass. Also called the **Fictive temperature.**

Strain-to-fracture   Elongation before fracture.

Strained-layer superlattice   An epitaxial thin film where the lattice spacing of the crystalline structure of the film material has been strained but not to the point of creating dislocations.

Stranski-Krastanov model (nucleation)   Nucleation on a surface that changes structure during the initial deposition.

Stress (adhesion)   A stimulus (mechanical, chemical, thermal, etc.) that tends to disrupt some feature or property of a film material, such as adhesion.

Stress, residual film (film formation)   The residual compressive or tensile stress in a film that results from the growth process, phase change during fabrication, or the differences in the coefficient of thermal expansion of the film and the substrate.

Stress corrosion   Chemical corrosion whose rate is enhanced by the presence of mechanical stress that is internal to the material or applied externally. See Wedging.

Stress tensor (adhesion)   The stress components of tension and shear that appear at the interface. If the material deforms or changes properties during the application of mechanical stress, the stress tensor may change.

Stress voiding (metallization)   The generation of internal voids by the movement of atoms under a tensile stress.

Striations (plasma)   Visual bands in the plasma that are due to plasma instabilities.

Strike (electroplating)   A thin (≈ 1 micron) electrodeposited film that is to be overlaid with other deposited materials. Also called a **Flash.**

Stringers (metallurgy)   A continuous filament of a sulfide found in steels. Stringers can lead to porosity and pitting corrosion. See Inclusions.

Stripe, conductor (electrical)   A thin film conductor line produced using masking or etching techniques.

Strippable coating (cleaning)   A liquid coating, such as a soap or a liquid polymer, that is applied to a surface and solidifies into a film that protects the surface from recontamination during some stage of processing. The coating material is removed during the subsequent cleaning processing.

Strippable coating, solid (cleaning)   A liquid coating applied to a surface that solidifies into a flexible film and whose purpose is to protect the surface from recontamination during some stage of processing. The strippable coating can also be used to coat over particles that are removed when the coating is removed. See Tack tape.

Stripping (cleaning)   The removal of a film, coating, or reaction layer from a surface.

Structure zone model (SZM) (film formation)   A diagram showing the morphology of a deposited film as a function of some deposition parameter. Examples: Temperature for vacuum evaporation; Gas pressure and temperature for sputter deposition. See Movchin-Demchishin diagram; Thornton diagram.

Styles of learning (manufacturing)   The way people learn. Some people are more receptive to visual information and some are more receptive to auditory information. To be most effective in transferring information both should be used. Important in operator training. See Technology transfer.

Styles of thinking (technology transfer)   The characteristic of the way that people think (**Synthesis, Realist, Idealist, Analyst, Pragmatist**). An important consideration in communication during technology transfer. See Technology transfer.

Sub-stoichiometric compound   A compound that does not have the correct ratio of elements to have the most stable structure. Examples: TiN_{1-x}; SiO_{2-x}. See Stoichiometric.

Sublimation (PVD technology)   Thermal vaporization from a solid surface. See Evaporation.

Sublimation pump (vacuum technology)   A capture (getter) pump in which the getter material is periodically renewed by sublimation from a solid source. Example: Titanium sublimation pump. See Vacuum pump.

Sublimation source (vaporization)   A vaporization source for heating materials, such as chromium, that sublimes rather than evaporates. The sublimation source can function best by ensuring good thermal contact between the heater and the solid. Examples: Electroplated chromium on a tungsten heater; Heating by radiation in an oven-like structure; Direct e-beam heating of the surface of the solid.

Substrate (PVD technology) Surface on which the film is being deposited. See Real surface.

Suceptor films Electrically conductive films deposited on paper or plastic packaging, which are heated by microwaves in order to make the package into an oven. Example: Microwave popcorn packaging.

Suck-back (vacuum technology) When the mechanical pumps stop, air will suck-back from the exhaust side to the low-pressure side, bringing with it oil contamination from the mechanical pump.

Suction The action of pushing a material toward a region of lower pressure. Generally by generating a vacuum so as to cause atmospheric pressure to push material toward the vacuum. Generally the vacuum used is very rough, such as a fraction of a psi. Example: Sucking liquid through a straw.

Sump (cleaning) The liquid reservoir into which condensed vapors drain. See Degreaser.

Superconductivity The disappearance of electrical resistance in a material below a certain temperature (critical temperature).

Supercritical fluid (SCF) (cleaning) A vapor that has been compressed to a pressure above its critical pressure and heated to above its critical temperature. In this condition the vapor and the liquid have indistinguishable properties.

Superhard materials Materials having a hardness greater than about 40 GPa. Examples: diamond = 100 GPa and cubic BN = 40 GPa.

Supersaturation The unstable condition when the vapor pressure of a material is above the saturation vapor pressure. Condensation is initiated by introducing condensation nuclei.

Suppliers (manufacturing) Organizations from outside the company that supply materials, piece-parts, equipment, etc. Also called **Qualified suppliers** if some basic criteria must be met.

Surface analysis (characterization) The determination of the chemical composition or the atomic arrangement of a surface or the near-surface region of a material.

Surface energy The energy associated with the non-symmetrical coordination of atoms in the surface. This energy determines the maximum size of a droplet, the maximum size of a void in a fluid, the wetting of a fluid on a surface, and the agglomeration of atoms on a surface. Measured in dyne/cm; ergs/cm^2; mJ/m^2.

Surface engineering Changing the properties of a surface to meet a specific requirement. This can be done by applying a film or coating to the surface to create a new surface (Overlay coating) or by changing the properties of the existing surface (Surface modification). Also called **Surface finishing**.

Surface enrichment The enrichment of some component of the bulk composition in the surface region as compared to the bulk. This may be due to loss of some constituent from the surface region or the preferential diffusion of species from the bulk to the surface region.

Surface finishing Changing the properties of a surface to meet a specific requirement. This can be done by applying a film or a coating to the surface to create a new surface (Overlay coating) or by changing the properties of the existing surface (Surface modification). Also called **Surface engineering**.

Surface mobility (adatom, film formation) The ability of a deposited atom (adatom) to move over the surface before it nucleates and becomes immobile.

Surface modification Changing the chemical, physical, mechanical, or morphological properties of a surface. Substrate material is present in the modified surface.

Surface, non-removable (vacuum technology) The surfaces in a vacuum chamber that cannot be removed for cleaning. Examples: Chamber walls; Feedthroughs; Tooling.

Surface, removable (vacuum technology) The surfaces in a vacuum chamber that can be removed for cleaning. Examples: Fixtures, Liners, Shields.

Surface roughness (substrate) The measure of the roughness of a surface from a mean value. See Roughness, surface (R_a).

Surface segregation Segregation of a material to the surface. Example: Diffusion of chromium through gold metallization to the surface where it oxidizes. The surface acts as a "Sink" for the chromium.

Surfactant (surface-active agent) (cleaning) A compound that reduces the surface tension between two fluids or between a fluid and a solid. Surfactants can be **high-foam** for use with static tank cleaning or **low-foam** for spray applications.

Susceptor, rf heating An electrically conductive material that can be heated by rf and it in turn can heat a material that is in contact with it. Carbon is often used as a susceptor material in PVD and CVD technology.

Synthesis reactions (CVD) Reactions involving two precursor species resulting in the deposition of a compound such as a metal carbide, oxide, nitride, etc.

T

Tack A measure of the "stickiness" of a surface.

Tack pad, floor (contamination control) A sticky (high-tack) surface placed on the floor and used to clean contamination from the soles of shoes and shoe coverings.

Tacky tape (cleaning) A sticky (**High-tack**) surface used to clean particulates from surfaces without leaving a significant amount of residual chemicals. See Strippable coating.

Tape test (adhesion)   A go or no-go (**Pass or Fail**) comparative adhesion test in which an adhesive tape is applied to a film surface, then rapidly pulled from the surface. Usually the film is scored under the area of test so that the tape pulls on a free edge of the film. See Adhesion tests.

Target (sputtering)   The surface being sputtered. Usually at a cathodic potential with respect to a plasma. Targets can be formed by machining, rolling, melting, vacuum melting, sintering, CVD, and plasma spraying.

Target, conformal (sputtering)   When the sputtering target is conformal with the substrate geometry. Example: Hemispherical target sputtering onto a hemispherical surface. See Fixtures.

Target, movable (sputtering)   A sputtering configuration where the sputtering target is moved while the substrate remains stationary. Used when coating very large substrates.

Target, opposing (sputtering)   When two or more (multiple of twos) planar unbalanced magnetrons face each other and the substrate is passed between the targets. The magnetic fields of the targets are such that the escaping magnetic field lines go from one target to another.

Target assembly, sputtering   The component of the sputter deposition system that contains the sputtering target, the target backing plate (if used), and the target cooling assembly. See Backing plate, target.

Target bonding (sputtering)   Joining the target to the backing plate with a high thermal conductivity bond. Bond can be inspected by thermal analysis or ultrasonic inspection. See Backing plate.

Target conditioning (sputtering)   Sputtering a target with a shutter closed or the substrates out of line-of-sight, to remove natural contamination layers such as oxides from the target surface.

Target poisoning   Reaction of the surface of a sputtering target either with the reactive gas being used for reactive deposition or with a contaminant gas. The reacted layer causes a change in the performance of the sputtering target.

Target shielding (sputtering)   Shielding of the target to prevent establishing a plasma between the shield and the target. See Paschen curve.

Tear resistance (web)   Resistance to tear as measured by ASTM 1004.

Technological surfaces   See Real surface.

Technology transfer   The transfer of a product design and fabrication technology from Research and Development (R&D) into Manufacturing. This includes issues dealing with manufacturability and scale-up as well as the ability of individuals to communicate with each other both through written (formal) documents such as specifications and through informal and formal personal interactions (e.g., meetings).

Tellurium breath (safety)   A bad case of halitosis resulting from breathing tellurium vapor.

Temperature   A measure of the average kinetic energy of particles in a material. It is important in communication between individuals that each person knows in what temperature units the other is using since normally the units are not specified. Example: "The substrate is heated to 100 degrees" (C or F?).

Temperature coefficient of resistance (TCR)   The rate of change of resistance with temperature. The change is positive for metals and negative for insulators and semiconductors.

Temperature scale, Centigrade (°C)   The temperature scale in which the freezing point of water is 0°C and the boiling point of water, under standard conditions, is 100°C. The degree centigrade has the same value as the degree Kelvin. Also called the **Celsius temperature scale.**

Temperature scale, Fahrenheit (°F)   The temperature scale based on the freezing point of water being 32°F and the boiling point of water under standard pressure conditions being 212°F.

Temperature scale, Kelvin (K)   The temperature scale where zero is the point of no atomic or molecular motion and the heat content of a material is zero. The Kelvin degree has the same magnitude as the Centigrade degree. Absolute zero is 0 K and -273.15°C.

Tempered (fully tempered) glass   Glass that has a high compressive stress on the surfaces and a high tensile stress at the midplane. When fractured, the tempered glass breaks up into small shards. Also called **Toughened glass.** See Hardened glass.

Tempering (glass)   To place the surface of the glass in compression by heating above the strain point, then quenching the surface region before the interior has a chance to cool, thus giving a higher fracture strength. See Tempered glass; Hardened glass.

Tempering (metal)   Heating briefly at a high temperature or heating at a low temperature to begin precipitation hardening and thereby creating a tougher material.

Tempering (substrate)   Removal of internal stresses by heating above the **Glass transition temperature** (glasses) or above the **Annealing temperature** (metals).

Tensile stress (thin film, PVD technology)   A stress resulting in the atoms being farther apart than they would be in a non-stressed condition. The tensile stress tries to make the film material contract in the plane of the film.

Terne (electroplating)   A lead-tin alloy electroplated coating on steel that is often used for roofing.

Terpene (cleaning) A natural homocyclic hydrocarbon solvent derived from plant life. Includes limonene, which is derived from citrus fruit, and pinene, which is derived from pine trees. Example: Turpentine.

Tesla (T) A unit of magnetic field intensity equal to 1 Weber/m². See Gauss.

Testing-to-a-limit (adhesion) Testing to a defined stress level. If the film does not fail it may be used as product. Example: Wire-pull test to a given load. See Adhesion.

Texture (crystalline) The preferential crystallographic orientation of lattice planes in a crystalline structure.

Texture (surface) The roughness, wave pattern, or other periodic morphological feature that describes a surface. See Orange peel; Capillary waviness.

Thermal control coating (window) A coating on windows that is used to reflect heat back into a room or keep out of a room. See Low-E coating; Solar control coating.

Thermal decomposition (CVD) The fragmentation of a molecule by heat alone.

Thermal desorption spectrum The species and amount of material desorbed as a function of temperature. This spectrum indicates how well the species is bonded or trapped in the solid.

Thermal gravimetric analysis (TGA) Chemical analysis by weight change as a function of temperature.

Thermal ionization Ionization in a high-temperature combustion flame. Also called **Flame ionization**.

Thermal oxidation Formation of an oxide surface layer by heating a surface in oxygen. Examples: Formation of a passive oxide on stainless steel by heating to 450°C in very dry (-100°C dew point) air; Oxidation of a clean silicon surface by **Rapid thermal processing (RTP)**.

Thermal spray processes A coating process where material (wire, rod, powder) is melted by a flame, plasma, electric arc, or some other means and the molten particles are propelled to the substrate surface in a high-velocity gas stream where they are splat cooled at a high quench rate. See Arc-wire spray; Plasma spray; Flame spray; High-velocity-oxygen-fuel (HVOF) spray.

Thermal strengthening Strengthening a high coefficient of expansion, low-thermal-conductivity material, such as glass, by putting the surface in compression by heating the material to above its strain point then rapidly cooling the surface to below the strain point so that when the interior cools it is placed in tension. This puts the surface region into compression.

Thermal stress adhesion test (adhesion) Subjecting a coating-substrate structure to an elevated temperature to introduce stress due to the differences in thermal coefficient of expansions of the materials. The stress may cause failure or may introduce flaws that cause failure in subsequent testing. See Adhesion tests.

Thermal vaporization (PVD technology) The vaporization of a material by raising its temperature. A useful vaporization rate for PVD processing is when the equilibrium vapor pressure is above about 2 mTorr. See Evaporation; Sublimation.

Thermalization (vacuum technology) The reduction of the energy of an energetic particle to the energy of the ambient particles by collision, as it passes through the ambient.

Thermionic emission Electron emission from a heated surface. This term is a misnomer since generally few ions are emitted from a heated surface for most materials. Exceptions are fluorine, cesium, potassium, and rubidium, which can be ionized by evaporation from a heated surface. See Thermoelectronic emission.

Thermistor gauge (vacuum technology) A form of the Pirani gauge in which the resistor element is a semiconductor material rather than a metal.

Thermocompression (TC) bonding The bonding of two surfaces under pressure and heat. Example: Thermocompression wire bonding of a wire to a metallized surface. See Ultrasonic bonding.

Thermocouple A temperature-measuring device consisting of two dissimilar metals joined together such that the voltage generated across the junction is dependent on the temperature of the junction.

Thermocouple gauge (vacuum technology) A pressure gauge that measures gas density by the cooling effect on a thermocouple junction. See Vacuum gauge.

Thermoelectronic emission Electron emission from a heated surface. Sometimes called **Thermionic emission**, which is poor terminology.

Thick film (PVD technology) A thick (> 0.5 micron) film deposited by PVD (or CVD) processing.

Thick film (hybrid microcircuits) A conductive or insulating coating prepared by painting, screenprinting, or dip coating a slurry onto a surface, followed by high-temperature firing to remove binders and fuse the material to the surface. Thick films can be used to form conductive, resistive or insulating layers or patterns. Patterns can be applied by **Screenprinting**.

Thickness, geometrical (film characterization) The film thickness as measured in units of length. Examples: Microns; Ångstroms; Mils; Nanometers.

Thickness, mass (film characterization) The film thickness as measured by mass per unit area. Example: Micrograms per square centimeter ($\mu g \cdot cm^{-2}$).

Thickness, optical (optical) The geometric thickness multiplied by the index of refraction.

Thickness, property (film characterization) The thickness measured by some property of the film, such as optical adsorption.

Thin film (PVD technology)   There is no universally accepted definition of the term "thin film." Generally the term is applied to deposits having a thickness of less than 0.5 micron. The term can be used to describe surface layers that affect the optical, electrical, or chemical properties of a surface; and in some cases the thin film affects the physical and mechanical properties of a surface, such as the abrasion resistance. See Thick film.

Threshold limit values (TLV) (safety)   The maximum amount of a chemical that a worker can be exposed to continuously or as a time-weighted average (TWA) as defined by OSHA. Examples: Trichloroethylene 270 mg/m^3; Arsine 0.05 mg/m^3; Chlorine 1 mg/m^3.

Throttling (vacuum technology)   Reducing the conductance of vacuum plumbing by reducing the crossectional area by use of a valve or an orifice.

Throughput, mass (vacuum technology)   The amount of gas measured in pressure-volume units (Torr-liters) flowing through the pump or the system per unit of time.

Throughput, product   The number of units per hour that are completely processed.

Throwing power (electroplating, PVD technology)   The ability of a deposition process to cover a rough surface or deposit material in high-aspect-ratio (depth-to-width) surface features such as vias.

Time-weighted average (TWA) (safety)   The amount of material in the air to which a worker can be exposed during an 8-hour shift (OSHA). See Permissible exposure limits (PEL); Short-term exposure limits (STEL).

Tin disease (electroplating)   The conversion of solid tin to a powder under low-temperature conditions.

Tin side (glass)   The side of the glass fabricated in a float glass plant that has been in contact with the molten tin. See Fire side.

Titration (chemical analysis)   The determination of the reactive capacity of a solution, such as acidity or alkalinity, by adding another solution with known composition, in known ratios, until a desired end point, such as color, is reached.

Tool (semiconductor processing)   System for performing a process (e.g., sputtering tool). Used synonymously with equipment. Also called a **Platform.**

Tool, wear-life of   How long a tool will perform satisfactorily. Measured as some tool function such as holes drilled, cut length, etc., under specified conditions.

Tooling   There is no universally accepted definition of the term "tooling" but it can be defined as the mechanical structure(s) in the deposition chamber that holds and moves the fixtures, vaporization source, shutters, masks, etc. Generally tooling is a non-removable structure in the system.

Tooling factor   The ratio of the observed condition, using sensors, during processing to the measured condition after processing. Example: Ratio of the film thickness on a quartz crystal monitor to the measured thickness of the film deposited on the substrate.

Top specks (float glass)   Particles of tin or its compounds found on the fire side of float glass.

Topcoat (PVD technology)   A film or coating that is put on a deposited film structure, generally by a separate process. Example: **Lacquer coating** on a deposited gold film to provide abrasion resistance.

Torr (or torr)   A unit of pressure defined as $1/760$ of a standard atmosphere. See Pressure, units of.

Total film stress   The total stress developed by the sum of the incremental residual film stresses in the film. Total film stress is a function of the film thickness. See Residual film stress.

Total life cost (equipment)   The installed cost plus the cost of operating and maintaining the equipment through its lifetime. See Installed cost; Cost-of-ownership (COO).

Total pressure   The sum of all the partial pressures of gases and vapors. See Dalton's Law of Partial Pressures.

Total pressure gauge   A vacuum gauge that measures the pressure effect of all gaseous and vapor species.

Toughness, fracture (adhesion)   The ability of a material to absorb energy and deform plastically before fracturing.

Toxic (chemical)   A chemical that has been shown to be toxic to mice. See Carcinogenic; Mutagenic.

Trace impurity   An impurity that occurs in a very small amount. Often in parts-per-million or parts-per-billion. See Minor impurity.

Trade-offs, design   Details of the design of a vacuum system that differ from the optimum vacuum design that are made to accommodate the use of the system in manufacturing. Examples: Large door openings to allow fixtures to be placed in the system; Side-pumped chambers to prevent items from falling into the pumping system (as can happen in a base-pumped system).

Trademark (™)   A letter, symbol, design, sound, etc., that has been registered with the U.S. Patent and Trademark Office and is used to establish an identity to a product or a producer.

Tradename (™)   A name given to a product or a process to establish an identity for the product or the process. Examples: C-Mag™ and Meta-Mode™ for PVD processing equipment; Viton™ and Nichrome™ for materials.

Training   Instruction of an operator in the proper procedures and techniques as defined by the Manufacturing process instructions (MPIs).

Training, formal (manufacturing)   Training in a classroom by experienced instructors. See Training, on-floor.

Training, on-floor (manufacturing) Training of an operator by having him/her work with an operator experienced in the process. Sometimes this is dangerous since bad habits can be passed from one to another. See Training, formal.

Tramp elements (electroplating) Undesirable ions in the electrolytic bath. See Dummying.

Transfer (printing) Metallization of a web of paper or plastic, cutting a pattern, and putting the pattern on a label, sign, etc.

Transit conductance (vacuum technology) Rate at which a specific molecule will go from one place to another in a specific geometry of surfaces.

Transition flow (vacuum technology) Gas flow conditions intermediate between viscous flow and molecular flow where the flow characteristics are determined by molecular collisions and collisions with the walls of the duct. Also called **Knudsen flow**.

Translucent Partially transparent, diffuse.

Transmission electron microscopy (TEM) An analytical technique that uses the scattering or diffraction of a high-energy electron beam as it passes through a thin film to image the microstructure of the film. Scanning transmission microscopy (STEM) is used to analyze a surface area.

Transverse direction (web coating) Direction normal to the direction that the web is moving. See Machine direction.

Trap (vacuum technology) A device for stopping or impeding the flow of vapors or particles through the system.

Traveler (manufacturing) Archival document that accompanies each batch of substrates detailing when the batch was processed and the specifications and MPIs used for processing. The traveler also includes the **Process sheet,** which details the process parameters of the deposition run.

Tribology Term coined by David Tabor from the Greek word "tribos," meaning rubbing. The science and technology of interacting surfaces in relative motion, and of associated subjects and practices such as lubrication, friction, and wear.

Trigger arc (arc vaporization) The high-voltage arc that is used to initiate the arc breakdown that is then sustained by the low-voltage, high-current arc.

Triode configuration (plasma) A plasma configuration where a plasma is established between a cathode and an anode, often with magnetic confinement, and ions are extracted out of the plasma to a third electrode, which is at a negative potential with respect to the plasma. Used in triode sputtering configurations.

Troy (t) weight scale Weight scale used for weighing precious materials where 1 pennyweight (dwt) = 1.54 grams, 1 troy ounce = 20 dwt or 30.8 grams, 12 oz (t) = 1 lb (t). Conversion: One oz (a) = 0.913 oz (t) and one lb (a) = 1.22 lb (t). See Avoirdupois weight scale.

Tuning (plasma) Matching the impedance of the load to that of the power supply so as to couple the maximum amount of energy into the load (plasma).

Turbomolecular pump (vacuum technology) A compression-type vacuum pump with a series of stator (stationary) and rotor (moving) blades that impart a change in velocity to the gas molecules by their being struck by the high-speed rotor blades and being reflected from the stator blades. The compression ratio that can be developed through the pump depends on the nature of the gas being pumped. Also called a **Turbo pump**. See Vacuum pump.

Turbulent flow A viscous flow with turbulent mixing. See Laminar flow.

U

U-value Heat transfer coefficient through a material or structure such as a double-glazed window, given in watts per meter square per degree Kelvin (W/m^2K).

Ultimate pressure (vacuum technology) The pressure in a system (or the inlet of a vacuum pump) toward which the pumping curve seems to be approaching asymptotically under normal pumping and processing conditions. Value will never be reached and depends on the sources of gases in the system. See Base pressure.

Ultrafine particle (cleaning) Particle having a diameter less than about 0.5 micron. Generally formed by vapor phase nucleation of vaporized material or the residue from the evaporation of an aerosol. Also called **Nanoparticle**. See Vapor phase nucleation; Gas evaporation; Nanophase materials.

Ultrahigh vacuum (UHV) (vacuum technology) The vacuum region where the pressure is less than about 10^{-8} Torr.

Ultrapure water (UPW) (cleaning) Water containing very low levels of ions, organic, particulate, and biological contamination. Specifications can be as stringent as: resistivity is 18 megohm-cm continuous at 25°C; particle count is less than 500 particles (0.5 micron or larger) per liter; bacteria count is less than one colony (cultured for 48 hours) per cc; and organics are less than one part per million (ppm). See Water.

Ultrasonic agitation (cleaning, electroplating) Agitation of a fluid, particularly in the boundary layer region, due to the formation and collapse of cavitation bubbles.

Ultrasonic bonding Bonding under pressure and ultrasonic "scrubbing." Example: Ultrasonic wire bonding to a film.

Ultrasonic cleaning (cleaning)   Cleaning due to the jetting action of the collapse of cavitation bubbles in contact with a surface.

Ultraviolet (UV) radiation   Electromagnetic radiation having a wavelength in the range of 0.004 to 0.4 micron. The short wavelength UV overlaps the long wavelength X-ray radiation and the long wavelengths approach the visible region.

Un-rack   To remove parts from a fixture. See Racking.

Unbalanced magnetron (sputtering)   A magnetron configuration in which the magnetic fields are arranged so as to allow some of the secondary electrons to escape from the vicinity of the cathode to establish a plasma between the target and the substrate. See Magnetron.

Undercuring (polymer)   When a polymer resin has not been fully cured, thereby leaving a large quantity of low-molecular-weight constituents in the polymer.

Underfiring, ceramic   When a sintered ceramic is not fired at a high enough temperature for a long enough time, producing a weak, porous, easily fractured material.

Unplasticized polyvinyl chloride (uPVC)   Polyvinyl chloride (PVC) that does not contain plasticizers that can migrate to the surface and become a source of contamination. Tubing of the material is used to distribute ultrapure water. Also used as a material in contact with a clean surface.

Uptime (vacuum technology)   The percentage of time in which the equipment is in condition to perform its intended function. See Downtime.

UV curable (polymer)   Polymer basecoat material that can be cured by exposure to ultraviolet radiation, thus avoiding the pollution problems associated with heat-curable polymers.

UV/Ozone (UV/O$_3$) cleaning   An oxidative cleaning process using ozone produced by ultraviolet radiation.

UV stabilizers   Chemical species added to polymers to adsorb the UV radiation and reduce decomposition of the polymer molecules by ultraviolet radiation.

Vacancy, lattice (crystallography)   A missing atom at a lattice site.

Vacuum   Pressure in a container that is less than the ambient pressure outside the container.

Vacuum, extra UHV (XUHV)   The pressure region less than about 10^{-10} Torr.

Vacuum, high   A gas pressure where there is molecular flow and a long mean free path for gas phase collisions. Generally taken as a pressure below about 10^{-5} Torr.

Vacuum, medium   The pressure range between rough vacuum and high vacuum.

Vacuum, rough   Pressure from atmospheric to about 50 mTorr. See Roughing.

Vacuum, ultrahigh (UHV)   The vacuum region where the pressure is less than about 10^{-8} Torr.

Vacuum arc remelting (VAR) (metallurgy)   Melting of an alloy by a low-voltage, high-current arc in a vacuum (10^{-3} mbar). See Vacuum induction melting; E-beam melting.

Vacuum breakdown (arc)   An arc between electrodes separated by a vacuum gap. See Flashover, surface; Arc, gaseous.

Vacuum cadmium plating   Vacuum deposition of cadmium on high-strength steel to avoid hydrogen embrittlement of the steel that can occur in electroplated cadmium. Also used to avoid water pollution problems. Also called **Vac cad plating**.

Vacuum chamber   The enclosure that is evacuated and in which the processing is to be performed. See Chamber, deposition.

Vacuum coating (paint industry)   A technique for pulling excess paint from a surface being painted, such as a continuous strip of molding moving at several hundred feet per minute.

Vacuum coating (thin film technology)   Coating (film) formed by depositing atoms or molecules in a vacuum environment. Coating process include both PVD and LP-CVD processes.

Vacuum-compatible materials (vacuum technology)   Materials that do not change characteristics in a vacuum and do not introduce contaminants into the system.

Vacuum deposition (surface engineering)   Any deposition process where a coating is applied in a chamber that is at less than ambient pressure. This includes PVD processes, some CVD processes, and some plasma spray processes.

Vacuum deposition (PVD technology)   Films deposited by thermal vaporization of a material in a vacuum so that particles that leave the source do not collide with gas molecules before they reach the substrate. Often used synonymously with **Vacuum evaporation**.

Vacuum distillation   Vaporization under reduced pressure to lower the boiling point of the material being vaporized and thus reduce the possibility of thermal degradation of the vaporized material.

Vacuum engineering   The design and construction of a vacuum system to meet processing requirements. This includes the design trade-offs that make the system more amenable to operation, cleaning, and maintenance. See Vacuum technology.

Vacuum evaporation (PVD technology, vacuum deposition processes) Thermal vaporization of a material in a vacuum so that particles that leave the source do not collide with gas molecules before they reach the substrate. Often used synonymously with **Vacuum deposition**.

Vacuum gauge A device for measuring gas pressures below the ambient atmospheric pressure. Often some property other than pressure is measured and related to the pressure by calibration.

Vacuum gauge, capacitance manometer gauge A vacuum gauge that uses the deflection of a diaphragm, as measured by the changing capacitance (distance) between surfaces, as an indicator of the pressure differential across the diaphragm—the pressure on one side being a known value.

Vacuum gauge, ionization, cold cathode gauge A vacuum gauge that uses ion current formed by electron-atom collisions as an indicator of the gas pressure (density). The electrons are formed as secondary electrons from ion bombardment.

Vacuum gauge, ionization, hot filament gauge A vacuum gauge that uses ion current formed by electron-atom collisions as an indicator of the gas pressure (density). The electrons are emitted from a hot thermoelectron emitting filament.

Vacuum gauge, Pirani gauge A vacuum gauge that uses the resistance of a heated resistor element, which changes due to gas cooling, in a Wheatstone bridge arrangement, as an indicator of the gas pressure (density).

Vacuum gauge, thermocouple gauge A vacuum gauge that uses the cooling of a heated thermocouple junction as an indicator of the gas pressure (density).

Vacuum gauge, viscosity gauge A vacuum gauge that uses the surface drag (deceleration) between the gas and a high-velocity surface to measure particle density. Example: Spinning rotor gauge. Also called a **Molecular drag gauge**.

Vacuum induction melting (VIM) (metallurgy) Melting of an alloy in vacuum (10^{-2} mbar) by induction heating. Alloys thus formed are often remelted using e-beam or arc melting. See Vacuum Arc Melting; E-beam melting.

Vacuum melting The melting of a metal in a vacuum environment to control contamination and aid in eliminating gaseous materials from the melt.

Vacuum oxygen decarburization (VOD) (metallurgy) Melting an alloy while blowing oxygen through the melt to lower the carbon content.

Vacuum processing Processes that are performed in the vacuum environment. Some vacuum processes are concerned with just lowering the ambient gas and vapor partial pressures to a level where the process can be performed in a satisfactory manner. Other vacuum processes use the vacuum environment to control the pressure level of special process gases.

Vacuum pump (vacuum technology) A device for reducing the gas pressure in a container to less than the ambient gas pressure. The vacuum pump can operate by capturing and holding the gases or by compressing and expelling the gases.

Vacuum pump, cryopump A capture-type pump that operates by condensation and/or adsorption on cold surfaces. Typically there are several stages of cold surfaces. Typically one of the stages will have a temperature below 120 K. See Cryosorption pump; Cryopanel.

Vacuum pump, diaphragm pump A compression-type vacuum pump that operates using a flexible diaphragm that changes the volume of the pumping chamber by mechanical motion. A very clean pump that can be exhausted to atmospheric pressure. Used to back a turbopump with a molecular drag stage.

Vacuum pump, diffusion pump (DP) A compression-type vacuum pump that operates by the collision of heavy vapor molecules with the gas molecules to be pumped, giving the gas molecules a preferential velocity toward the high pressure stages of the pump. Also called a **Diff pump**; **Vapor jet pump**.

Vacuum pump, getter pump A capture-type vacuum pump that operates by reaction of a surface with the gaseous species to form a non-volatile reaction product or by absorbing the gases into the bulk of the getter material. In reaction-type getter pumps the getter materials are often deposited by evaporation or sublimation. Adsorption-type getter pumps are sometimes called **Non-evaporative getter pumps**. See Sublimation pump; Ion pump; Getter.

Vacuum pump, ion pump A capture-type vacuum pump where a getter material is deposited by sputtering and gaseous ions are accelerated to the reactive surface to react with the surface or be physically buried in the depositing material. Also called a **Getter ion pump**.

Vacuum pump, mechanical pump A compression-type vacuum pump with moving parts. The term is generally applied to pumps used for roughing or backing (e.g., oil-sealed mechanical pump, piston pump, diaphragm pump, etc.) and not high-vacuum pumps (e.g., turbomolecular pumps).

Vacuum pump, sorption pump A capture-type vacuum pump that operates by cryocondensation of gases on a large-adsorption-area, cryogenically cooled (< -150°C) surfaces.

Vacuum pump, sublimation pump A getter pump where the getter material, such as titanium, is deposited by sublimation from a solid surface.

Vacuum pump, turbopump A compression-type vacuum pump with a series of stator (stationary) and rotor (moving) blades that impart a change in velocity to the gas molecules by their being struck by the high-speed rotor blades and being reflected from the stator blades. The compression ratio that can be developed through the pump depends on the nature of the gas being pumped.

Vacuum pump, Venturi A vacuum pump using the Venturi effect (i.e., entrainment of gas in a high-velocity gas or liquid stream) to reach a vacuum as low as 60 Torr.

Vacuum surface (vacuum technology) A surface in contact with the vacuum environment.

Vacuum surface, non-removable The surfaces, such as chamber walls, that are not easily removed from the system and must be cleaned in place.

Vacuum surface, removable Surfaces, such as fixtures, that are routinely removed from the system or surfaces, such as liners, that can be removed from the system for cleaning.

Vacuum technology (vacuum technology) The operation, cleaning, maintenance, and repair of a vacuum system so that it continues to meet processing requirements. See Vacuum engineering.

Vacuum zoning (vacuum technology) Dividing a vacuum chamber into separately pumped regions connected by low conductance openings.

Vacuum-based ion plating (film deposition) A special case of ion plating where the deposition is done in a high vacuum and the concurrent or periodic bombardment is provided by ions accelerated from an ion gun or a plasma source. Also called **Ion beam assisted deposition (IBAD)**.

Valence The number of excess charges (positive or negative) associated with an atom, a molecule, or a radical species.

Valve (vacuum technology) A mechanical device that can start, stop, or regulate the flow of a gas or fluid by use of a moving part that opens or obstructs a passage.

Valve, angle A valve that does not provide an optically straight path through the valve opening.

Valve, ballast (vacuum technology) A valve to allow the admittance of a dry gas into the foreline or the inlet of the roughing pump to prevent condensation of vapors by compression and prevent suck-back. See Suck-back.

Valve, butterfly (vacuum technology) A valve in a tube that operates by rotating a plate along an axis that is through the diameter of the tube.

Valve, check A valve that limits flow to one direction only.

Valve, external A valve that seals at a high pressure differential such as between atmospheric pressure and vacuum. See Valve, isolation; Valve, internal.

Valve, gate (vacuum technology) A large-area, high-conductance valve that is sealed by a moving plate.

Valve, high-vacuum (vacuum technology) The high conductance valve between the high-vacuum pump and the volume to be evacuated.

Valve, in-line A valve that provides an optically straight path through the valve opening.

Valve, internal A valve that is internal to the vacuum system that seals with a low pressure differential. Also called isolation valve. See Valve, external.

Valve, isolation (vacuum technology) A valve that isolates one vacuum chamber from another at a low pressure differential. Also called an internal valve.

Valve, manual A hand-operated valve.

Valve, normally closed (NC) A valve that is closed when there is no actuating force.

Valve, normally open (NO) A valve that is open when there is no actuating force.

Valve, pendulum (vacuum technology) A high-conductance valve that is sealed by a moving plate. Typically after breaking the seal the plate will move in a circular (swinging) direction to uncover the valve opening.

Valve, pipeline See Valve, poppet.

Valve, poppet (vacuum technology) A mechanical vacuum sealing valve where the motion of the sealing plate is normal to the plane of the seal. Also called a **Pipeline valve** (Europe).

Valve, pneumatic A valve that is actuated by a piston driven by air pressure. See Valve, solenoid.

Valve, pressure relief A safety valve that opens when the pressure on the system-side exceeds some predetermined value.

Valve, roller (vacuum technology) A valve used on a continuous sheet where rollers contact the surface above and below the sheet to give vacuum sealing.

Valve, roughing (vacuum technology) The valve between the roughing pump and the volume to be evacuated.

Valve, slit (vacuum technology) A mechanical sealing valve that has a long, narrow, rectangular opening. Often used in passing flat and thin substrates, such as silicon wafers and architectural glass plates, into and out of a chamber. Also called a **Slot valve**.

Valve, soft-roughing (vacuum technology) A valve whose variable conductance allows the system to be rough-pumped slowly to minimize turbulence in the system.

Valve, soft-vent (vacuum technology) A valve whose variable conductance allows the system to be returned to ambient pressure slowly to minimize turbulence in the system.

Valve, solenoid A valve that is actuated by an electric solenoid. See Valve, pneumatic.

Valve, throttling (vacuum technology) A valve used to control the pumping speed. It may be a special valve (such as a variable conductance valve), or it may be a standard open-closed valve that is only closed partway, possibly by hand.

Valve, variable conductance (vacuum technology) A valve whose conductance can be varied in a controlled manner. Generally refers to large valve openings. See Valve, variable leak.

Valve, variable leak (vacuum technology) A variable conductance valve that is designed to control a very small gas flow (leak) of a few to several hundred standard cubic centimeters per minute (sccm). Used as part of a mass flow controller. See Mass flow controller.

Valve, vent (vacuum technology) Valve to allow the system to be returned to ambient pressure. See Valve, soft-vent.

Valve metals Metals that can be oxidized to form dense, coherent oxide layers on the surface. Examples: Aluminum; Titanium; Niobium.

Vapor A gas that is easily condensed by cooling, compression, etc. The term gas is often used in a context that includes vapors. Also called **Vapour**. See Gas.

Vapor cleaning Cleaning by the condensation of a solvent vapor on a cold surface above a hot liquid sump. The condensed solvent and contamination flows off into the sump. Cleaning continues until the part reaches the temperature of the vapor and condensation ceases. Also called **Vapor degreasing.**

Vapor condensation (cleaning) A solvent is heated to form a vapor cloud above the surface and a cold part is suspended in the vapor. The solvent condenses on the part and flows off into the "sump," carrying the contaminants with it. When the part is heated to the vapor temperature, condensation ceases and cleaning action stops. See Degreaser, vapor.

Vapor dry (cleaning) Drying by condensing a vapor on the surface that displaces the water and flows off into a "sump." Usually done hot so that the high-vapor-pressure drying agent is rapidly vaporized from the surface.

Vapor forming (PVD, CVD) Fabrication of a free-standing structure by depositing material from a vapor (CVD or PVD) onto a mandrel, then removing the mandrel. See Electroforming.

Vapor jet pump (vacuum technology) A kinetic pump where the gas molecules are entrained in a jet of fluid vapor. See Diffusion pump; Venturi tube.

Vapor lock Interruption of fluid flow through a channel by the creation of a vapor bubble in the channel due to excessive heating.

Vapor phase epitaxy (VPE) (PVD technology) Formation of single crystal films by Chemical vapor deposition (CVD) processes. See Chemical vapor deposition (CVD).

Vapor phase etching (cleaning) Chemical etching using a vapor instead of a fluid.

Vapor phase nucleation The development (condensation) of nuclei in the gas phase due to multi-body collisions. See Ultrafine particles; Gas evaporation; Black sooty crap (BSC).

Vapor pressure, equilibrium The pressure of the vapor of a solid or liquid above the surface in a closed container such that as many particles return to the surface as leave the surface. Also called the **Saturation vapor pressure**.

Vapor pump, cryopanel A capture-type vapor pump that removes vapors, by cryocondensation on large-area surfaces, which are at temperatures of -100 to -150°C. The surfaces may be in the vacuum chamber where there is high conductance to the surfaces. At this temperature the vapor pressure of water is very low. Also called a **Meissner trap**.

Vaporization (volatilization) The conversion of a solid or a liquid to a vapor by any means such as thermal, arcing, sputtering, etc.

Vapour See Vapor. British spelling for vapor.

Velocity A vector quantity of motion that has both speed (a scalar, not a vector, quantity) and direction. Example: Miles-per-hour is a speed while miles-per-hour to the north is a vector. Note: A change in direction is a change in velocity even when there is no change in speed.

Vented bolt (vacuum technology) A bolt that has a hole down the axis or a groove through the threads to avoid forming a virtual leak when the bolt is inserted in a blind hole.

Venting (vacuum technology) Bringing the system up to ambient pressure. Also called **Backfill**.

Venturi tube (vacuum technology) A constriction in a pipe that causes an increase in the velocity of a fluid or a gas and creates a vacuum that can be used to draw fluid or gas into the main flow of fluid or gas through a port in the constriction. Examples: Used in a carburetor to draw fuel into the air stream; Used as a suction pump in a chemistry laboratory when placed on a water faucet.

Very low pressure CVD (VLP-PECVD) Plasma enhanced CVD at a pressure where ions can be accelerated to appreciable energies (< 15 mTorr). Often used in conjunction with a PVD process to give a hybrid process such as depositing a metal carbide by sputtering the metal and obtaining the carbon by PECVD from C_2H_2.

Via (semiconductor processing) A hole that extends from one level of a multi-layer structure, through an intermediate layer, to another layer. To make electrical contact between layers the hole must be filled with a conductive material.

Vibration (vacuum technology) Repetitive movement of surfaces (cm/s) that can contribute to particulate generation by wear and pinhole flaking. Origin of vibration may be the system location or from the mechanical pumping system.

Vicinal surface   A surface of a single crystal material that has been cut and polished at an angle to a crystallographic plane in order to give a "stepped-surface" on an atomic scale. Also called an **Off-cut surface**.

Vickers hardness number   The expression derived from the force used and the projected area of an imprint obtained by a square shaped (ASTM E 348) diamond indenter forced into a surface. Abbreviated HV (formally VHN). $HV = 1854.4 \ P/d^2$ where P = grams force and d = length of diagonal in microns.

Virtual leak (vacuum technology)   A conduction path from an internal trapped volume to the main volume of a vacuum system (no connection to the outside ambient environment). Example: Void below the bolt in a blind, tapped hole.

Viscous flow (vacuum technology)   Gas flow where the mean free path for collision is very small compared to the dimensions of the system. Viscous flow may be laminar or turbulent.

Visible radiation   Electromagnetic radiation that is visible to the human eye. Electromagnetic radiation in the wavelength range of ~0.38 to ~0.78 micron.

Vitreous (material)   A glassy (no discernible crystal structure) material. See Crystalline.

Void (film growth)   A region lacking solid matter. The void may be internal with no connection to a free surface or it can be connected to a free surface. Also called a **Pore**, particularly if elongated.

Volatile organic compounds (VOCs) (cleaning)   Organic compounds, such as solvents, that have boiling points below 138°C.

Volcanoes (film characterization)   Eruptions in a film where reactive gases or vapors have reacted with the underlying material to form corrosion products.

Volt (electric)   The unit of potential difference or electromotive force, between two points, in the MKS system of units for which one coulomb of electric charge will do one joule of work in going from one point to the other under 1 volt potential.

Voltage (electricity)   The electrical potential difference between two points. See Volt; Potential.

Voltage polarity (electricity)   The indication of which direction an electron will flow between two points due to the voltage between the points. For example, electrons will flow away from the negative pole toward a positive pole or an electron will flow from a more negative potential to a less negative potential.

Volume flow rate (vacuum technology)   The volume of gas passing through a pump or system at a specific pressure and temperature. Measured in liters/sec. Also called **Throughput**. See Mass flow; Pumping speed.

Volume gas (vacuum technology)   The gas contained in an enclosed volume and that is free to move about. Does not include gas from other sources such as outgassing, virtual leaks, etc.

Vycor™ (glassware)   A 96% fused silica glass that is more formable than pure fused silica.

Wafer (semiconductor processing)   A specific type of substrate, usually a thin disk of silicon or GaAs.

Wall creep (vacuum technology)   Movement of an adsorbate along a surface.

Warm-edge technology (energy conservation)   Methods of separating two glass sheets (window pane) around the edge using a medium that has low thermal-transport properties.

Warm-up time (diffusion pump)   The time necessary to bring the pumping fluid in a diffusion pump up to the proper operating temperature.

Warm-up time (mass flow meter)   The time for a mass flow meter to warm to operating temperature.

Water (H_2O) (cleaning)   Common cleaning solvent for ionic contaminants and as a rinsing agent. Often used as a water-alcohol mixture to lower the surface energy of the fluid.

Water, de-ionized (DI water)   Water with a low ionic content as measured by its electrical conductivity. Sometimes used synonymously (incorrectly) with Pure water or Ultrapure water. This is incorrect since DI water can still have organic and particulate contamination.

Water, distilled   Water purified by distillation, often by several distillation stages (i.e., triply distilled water).

Water, hard   Water containing a high concentration of easily precipitated ions such as Ca, Mg, Fe.

Water, pure (cleaning)   Water purified by reverse osmosis (RO) along with carbon filtration (organics) and mechanical filtration (particulates). Often used as a solvent and as a rinsing agent where requirements allow.

Water, semiconductor grade (cleaning)   Water that is pure enough to meet the requirements of the semiconductor processing industry. See Water, ultrapure.

Water, soft   Water in which easily precipitated ions, such as Ca, Mg, Fe, have been replaced by more soluble ions such as Na. Often used as an intermediate rinsing agent.

Water, ultrapure (cleaning)   Water containing a very low concentration of materials (ionic, organic, inorganic, biological, etc.) that will leave a residue on evaporation. Produced by ion exchange (ionic species), carbon filtration (organics), and mechanical filtration (particles, biological agents). The most pure rinsing material.

Water adsorption (outgassing) Amount of water taken up by a material after a 24-hour immersion.

Water jet pump A kinetic vacuum pump where the gases are entrained in a jet of water. See Venturi tube.

Water spot (cleaning) The spot of residue left from the evaporation of impure water.

Water vapor transmission rate (WVTR) (film characterization) The amount of water vapor transmitted thorough a film in units of amount per unit area per unit time. Units of g/m2-d. Also called **Moisture transmission rate (MTR)**; **Water vapor permeability.**

Water-break test (cleaning) A test for hydrophobic contamination by observing the sheeting action of water on the surface. If the sheet of water avoids certain areas, hydrophobic contamination is to be expected in those areas. See Sheeting.

Watt (W) The SI unit of power.

Wavelength The distance between two points having the same phase in two consecutive cycles of a periodic wave. Example: The wavelength of electromagnetic radiation.

Wavenumber The reciprocal of the wavelength.

Weak surface layer (adhesion) When the surface layer is weak either due to a low-molecular weight layer (polymer) or surface flaws (brittle solid). During film deposition this weak region becomes part of the interphase material, resulting in poor apparent adhesion.

Wear The removal of material by friction between materials in moving contact.

Web (PVD technology) A thin, flexible membrane that may be solid or perforated.

Web, polymer, properties of Some properties of web materials are: surface energy, tear strength, puncture resistance, impact strength, clarity, flexibility, heat-sealing characteristics, thermal stability, shrink-film performance.

Web coating (PVD technology) Depositing a film on a web of material, usually of a polymer or paper. Aluminum is a commonly deposited film material in web coating.

Wedging action (corrosion, adhesion) When there is corrosion at an interface and the solid or gaseous corrosion product expands and exerts a stress on the interface, thus enhancing the corrosion rate and the loss of film adhesion.

Weight A measure of the gravitational attraction of a body. Often used synonymously (but incorrectly) with mass. See Mass.

Weight percent (alloy) The percentage by weight of one material in an alloy composition. Abbreviated wt%. See Atomic percent. Example: An alloy of W:10wt%Ti has the same composition as a W:30at%Ti alloy.

Weight-gain analysis (cleaning) An analysis of the rate and amount of material absorbed by another material in a given environment. Example: This is of concern in recontamination, where moisture pickup after bakeout will determine the storage environment needed.

Weight-loss analysis (cleaning) The analysis of the desorption or extraction of material from the bulk as a function of the ambient environment. Example: Desorption of water during vacuum baking will determine the time and temperature needed to desorb the water to an acceptable level.

Welding (fusion weld) (vacuum technology) Joining two materials by melting and mixing the materials in the interfacial region. Care must be taken that flaws and stresses are not generated in the heat affected zone (HAZ) during cool-down.

Wetting agent (cleaning) A chemical that reduces the surface energy of a fluid, which makes it flow over a surface (wet) more easily. Example: Alcohol in water.

Wetting angle (surface characterization) The angle that a drop of liquid makes with a surface as measured through the liquid.

Wetting growth (film formation) The lateral growth of nuclei on a surface due to the strong interaction of the adatoms with the surface. See De-wetting growth.

What-if game (vacuum technology) The question and answer session used to establish a fail-safe design by asking what will happen if something fails. Example: What happens if the power goes off for one minute? For an hour?

White metal White-colored metals such as aluminum, magnesium, antimony, or zinc and some of their alloys. Also called a **Gray metal**.

White room A clean area (about a Class 10,000) that uses many of the construction practices, equipment, and techniques of a cleanroom but does not use high volumes of filtered air. See Cleanroom.

Window (vacuum technology) A feedthrough that allows optical (optical window) or magnetic (magnetic window) radiation to pass through the chamber wall. See Feedthrough.

Window film A coated flexible polymer film that can be glued to a window for thermal control, either by reducing solar isolation or by radiant heat loss.

Window, process parameter The region between the process parameter limits that allows a satisfactory product to be produced. The larger the window, the more robust is the process. Example: 100°C ± 10°C - the window in 20°C.

Wipe clean (cleaning) See Wipe-down.

Wipe-down (cleaning, vacuum technology) Cleaning/drying a surface by wiping with a lint-free, compliant, low-extractable material, such as a cloth or sponge that contains a cleaning/drying fluid such as anhydrous alcohol.

Wire bond   An electrical connection to a surface made by pressing a section of wire under heat and pressure (thermocompression bonding) or pressure and ultrasonic scrubbing (ultrasonic bonding) against the surface. See Ball bond.

Witness plate (characterization)   A substrate that is not a part of the production batch but is used for characterizing some portion of the process or some film property such as film thickness, film stress, film adhesion, etc. Also called a **Monitor plate.**

Wolfrum   Another name for the element tungsten.

Workhardening   The hardening of a metal by repeated deformation, creating a lot of lattice dislocations. The hardening can be removed by annealing above the recrystallization temperature. Example: The knife-edge on a CF flange is hard because of workhardening of the stainless steel during machining. If the flange is heated to above 450°C, the knife-edge will be ruined by annealing.

Work of adhesion (W$_a$), thermodynamic (adhesion)   The thermodynamic work of adhesion (W$_a$) between two polymer materials (1 and 2), in ideal contact, is given by the **Dupre equation**: $W_a = \delta_1 + \delta_2 - \delta_{1,2}$ Where δ_1 and δ_2 are the surface energies and $\delta_{1,2}$ is the interfacial energy.

X

X-ray   Short wavelength (< 1 Å) electromagnetic radiation.

X-ray diffraction   (characterization) Diffraction, usually of crystalline lattices, using X-ray radiation.

X-ray emission (characterization)   The generation of characteristic X-rays by bombarding a surface with high-energy electrons.

X-ray fluorescence (XRF) spectroscopy (characterization)   The generation of characteristic X-ray radiation from a surface by bombarding the surface with X-rays. The emitted characteristic X-rays are characterized by their wavelength or their energy. The analytical technique is a non-destructive technique for determining element composition of a layer up to several microns in thickness, depending on the mass of the elements. See Micro X-ray analysis.

X-ray fluorescence (XRF) spectroscopy, energy dispersive (ED-XRF) (characterization)   X-ray fluorescence analysis (X-rays as the probing species and X-rays as the detected species) where the detected species is energy analyzed in a lithium-drifted detector. Used to determine the chemical composition of the material being probed or the film thickness when the X-rays can penetrate through the film.

X-ray fluorescence (XRF) spectroscopy, wavelength dispersive (WD-XRF) (characterization)   X-ray fluorescence analysis (X-rays as the probing species and X-rays as the detected species) where the detected species is analyzed as to its wavelength using a crystal diffractometer.

X-ray photoelectron spectroscopy (XPS) (characterization)   A surface analytical technique where the probing species are X-rays and the detected species are photoelectrons. The technique allows both species identification and the chemical bonding energy. See also Electron spectroscopy for chemical analysis.

X-ray thickness measurement   Measurement of the thickness (mass per unit area) by the attenuation of X-rays passing through the material. See Thickness.

Xerography   Xerography is done by the attraction of electrically charged "toner" to a paper that has been charged by exposure to an optical image, then fusing the toner to the paper with heat. The charging drum is coated with selenium or other photosensitive material. Also called **Electrography**.

Yield, product   The percentage of substrates that enter the production processing sequence that result in good product.

Yield, secondary electron   The number of electrons emitted from a surface per incident electron or incident ion. The secondary electron yield for electrons is much higher than for ions.

Yield, sputtering   The number of ejected (sputtered) surface atoms per incident high-energy bombarding particle (ion).

Yield stress   The lowest stress at which a material will begin to plastically deform under mechanical stress.

Young's modulus   The ratio of the applied tensile stress to the resulting elastic strain.

Zeolite   A high-surface-area mineral that is used in sorption pumps and cryosorption traps. The zeolite structure is characterized by internal cavities that are accessible through "windows" whose opening can be adjusted between 3Å and 7Å, which allows selective adsorption of gases or vapors depending on their molecular diameter. Zeolites must be regenerated at 200°C or more. Zeolites are also called **Molecular sieves**.

Zeroing (a meter)   Moving the indicator point of a meter reading to a zero reading at zero input.

Zeta potential   The electrical potential that exists across the interface between a solid and a liquid.

Printed and bound by CPI Group (UK) Ltd, Croydon, CR0 4YY

03/10/2024

01040338-0020